T0136466

THE INTELLECTUAL PROPERTY–REGULATORY COMPLEX

THE INTELLECTUAL PROPERTY–REGULATORY COMPLEX

OVERCOMING BARRIERS TO INNOVATION IN AGRICULTURAL GENOMICS

Edited by Emily Marden,
R. Nelson Godfrey, and
Rachael Manion

UBCPress · Vancouver · Toronto

25 24 23 22 21 20 19 18 17 16   5 4 3 2 1

Printed in Canada on FSC-certified ancient-forest-free paper
(100% post-consumer recycled) that is processed chlorine- and acid-free.

Library and Archives Canada Cataloguing in Publication

The intellectual property-regulatory complex : overcoming barriers to innovation in agricultural genomics / edited by Emily Marden, R. Nelson Godfrey, and Rachael Manion.

Includes bibliographical references and index.
Issued in print and electronic formats.
ISBN 978-0-7748-3178-9 (bound). – ISBN 978-0-7748-3180-2 (pdf)
ISBN 978-0-7748-3181-9 (epub). – ISBN 978-0-7748-3182-6 (mobi)

1. Intellectual property. 2. Agricultural biotechnology – Law and legislation.
3. Genomics. I. Marden, Emily, author, editor II. Godfrey, R. Nelson, author, editor
III. Manion, Rachael, editor

K1519.B54I58 2016   346.04'8   C2015-907446-0
                                                      C2015-907447-9

Canadä

UBC Press gratefully acknowledges the financial support for our publishing program of the Government of Canada (through the Canada Book Fund), the Canada Council for the Arts, and the British Columbia Arts Council.

UBC Press
The University of British Columbia
2029 West Mall
Vancouver, BC V6T 1Z2
www.ubcpress.ca

Dedication

From Emily Marden:

"To Roger, Elena, and Andreas, for their boundless support."

From R. Nelson Godfrey:

"Thanks Mom, for everything."

From Rachael Manion:

"To my family, with thanks."

Contents

List of Tables and Figures / ix

Foreword / x
HANNES DEMPEWOLF

Abbreviations / xii

Acknowledgments / xv

Introduction / 3
EMILY MARDEN, R. NELSON GODFREY, AND RACHAEL MANION

Part 1   Perspectives on Regulatory Regimes / 13

1   Biosafety and Intellectual Property Regimes as Elements of
the IP–Regulatory Complex: The Case of Canadian Sunflower
Genomics / 15
EMILY MARDEN, R. NELSON GODFREY, MATTHEW R. VOELL,
AND LOREN H. RIESEBERG

2   The Treatment of Social and Ethical Concerns in
Regulatory Responses to Agricultural Biotechnology:
A Historical Analysis / 42
SARAH HARTLEY

3 How the IP–Regulatory Complex Affects Incentives to Develop
Socially Beneficial Products from Agricultural Genomics / 68
GREGORY GRAFF AND DAVID ZILBERMAN

4 Stealth Seeds: Bioproperty, Biosafety, Biopolitics / 102
RONALD J. HERRING

**Part 2 Intellectual Property Mechanisms and
Counter-Movements / 141**

5 Implementing the *International Treaty on Plant Genetic
Resources for Food and Agriculture:* A Regulatory and Intellectual
Property Outlook / 143
CHIDI OGUAMANAM

6 Intellectual Property Management and Legitimization Processes
in International and Controversial Environments / 171
JEREMY HALL, STELVIA MATOS, AND VERNON BACHOR

Part 3 Future Directions / 193

7 A Governance Approach to the Agricultural Genomics
Intellectual Property–Regulatory Complex / 195
REGIANE GARCIA

8 Constructing an International Intellectual Property *Acquis*
for the Agricultural Sciences / 211
ROCHELLE COOPER DREYFUSS

List of Contributors / 234

Index / 239

Table and Figures

Table

6.1   Interview subjects by stakeholder category / 179

Figures

3.1   The apportionment, among different stakeholder groups in the US corn seed industry, of benefits and costs of *Bt* corn / 73

3.2   Interaction among different elements of the IP–Regulatory Complex / 80

4.1   Biosafety regime scenarios and social costs / 125

Foreword

HANNES DEMPEWOLF

The Food and Agriculture Organization of the United Nations estimates that 1 billion people go to bed hungry every night, almost 98 percent of them in the developing world. Climate change and population growth are predicted to make the situation much worse in the coming decades, but the crisis is already here. One billion hungry people is not acceptable: we must breed more nutritious, more resilient, and higher-yielding crops.

Of course, better seeds are not the only solution to world hunger, but they are certainly one of our key weapons in the battle for a more food-secure planet. Quantum leaps are required to enable farmers to grow more food on less land, with fewer inputs, and in increasingly challenging environmental conditions. And they have to occur quickly. Plant breeding needs to be accelerated, more efficient, and cheaper. Innovative solutions are clearly required.

Like almost every other area in the life sciences, crop improvement is benefiting from the genomics revolution. Many in the agricultural research community believe that the latest large-scale, high-throughput genomics and phenomics approaches can deliver faster, better-targeted crop improvement. Much-needed innovations in crop improvement are set to emerge, many of them based on the principle of applying "omics" techniques to plant breeding. Decades' worth of basic genomics research are nearing fruition at last.

Can it happen? What will it take? Certainly, and inevitably, more funding is needed. Sequencing a genome is the beginning, not the end. But one should not underestimate the crucial importance of appropriate intellectual property and regulatory mechanisms that encourage innovation and truly unleash the enormous potential of these technologies and make them accessible to those who need them most.

This is a significant challenge. Fortunately, there are useful precedents in the field of plant genetic resources, the biological raw material that provides the fundamental basis for any innovation in agricultural genomics. The global plant genetic resources community has come together over the past decade through the International Treaty on Plant Genetic Resources for Food and Agriculture (ITPGRFA), which now has an impressive list of 136 contracting party countries. International recognition of the interdependence of countries on each other's genetic resources for food and agriculture has led to the establishment of a multilateral system for access and benefit sharing. I hope the same spirit that led to the ITPGRFA will also guide decision makers when developing and negotiating the policies that will enable agricultural genomics to fulfill its potential in farmers' fields and orchards.

At the same time, we cannot forget that the only reason a genomics revolution in crop improvement is even possible is that we stand on the shoulders of proverbial giants: the farmers under whose watchful eyes the diversity of our crops has evolved, as well as the researchers, gene bank managers, and policy makers who worked tirelessly to preserve and utilize the world's plant genetic resources in the past. The contribution of all has to be recognized.

This edited volume comes at a crucial time. Every day we see huge genomics datasets being published under a bewildering array of different (and all too often unclear or contradictory) access regimes, with all too little thought given to linking the data in a clear and permanent way to seeds or plants in a gene bank. On the other hand, policy makers and regulators are struggling to keep pace with rapid technological advances in genomics, and therefore, perhaps understandably, all too often fail to agree on appropriate rules to effectively govern access to the data deluge. We want to encourage continued innovation but at the same time ensure that data and germplasm can actually be used to improve crops. How can we ensure maximum benefit to people in the developing world while simultaneously securing much-needed revenue from newly developed materials? It is clear that a focused discourse on these issues is necessary at both the national and international levels. The editors and authors of this volume have made a truly remarkable contribution toward this goal. I congratulate them on their important achievement and hope that policy makers and regulators will take note.

Abbreviations

ACTA	*Anti-Counterfeiting Trade Agreement*
APEC	Asia-Pacific Economic Cooperation
BJP	Bharatiya Janata Party (Indian People's Party)
BSF	Benefit Sharing Fund (under the ITPGRFA)
CBAC	Canadian Biotechnology Advisory Committee
CBD	*Convention on Biological Diversity*
CEPA	*Canadian Environmental Protection Act*
CFIA	Canadian Food Inspection Agency
CGIAR	Consultative Group for International Agricultural Research
CIDA	Canadian International Development Agency
CIMMYT	International Maize and Wheat Improvement Center (Centro Internacional de Mejoramiento de Maíz y Trigo)
CIS	Common Implementation Strategy (European Union)
CJEU	Court of Justice of the European Union
CPGRFA	Commission on Plant Genetic Resources for Food and Agriculture
CTNBio	Comissão Técnica Nacional de Biossegurança (National Technical Biosafety Commission, Brazil)
DKSS	Deshiya Karshaka Samrakshana Samithi (National Agriculturalist Protection Committee, India)

EMBRAPA	Empresa Brasileira de Pesquisa Agropecuária (Brazilian Enterprise for Agricultural Research)
FAO	Food and Agriculture Organization of the United Nations
GATT	*General Agreement on Tariffs and Trade*
GEAC	Genetic Engineering Approval Committee (India)
GM	genetically modified
GMO	genetically modified organism
GURT	gene use–restriction technologies
HLRT	High Level Round Table on the Treaty (i.e., the ITPGRFA)
HP	HarvestPlus
IA	intellectual assets
IARC	International Agricultural Research Centre
IBAMA	Instituto Brasileiro do Meio Ambiente e dos Recursos Naturais Renováveis (Brazilian Institute for the Environment and Renewable Natural Resources)
ICAR	Indian Council for Agricultural Research
ICRISAT	International Crops Research Institute for the Semi-Arid Tropics
IDEC	Instituto Brasileiro de Defesa do Consumidor (Institute for Consumer Defense, Brazil)
IFPRI	International Food Policy Research Institute
ILCs	indigenous and local communities
INASE	Instituto Nacional de Semillas (National Seed Institute, Argentina)
IP	intellectual property
IPM	integrated pest management
ITPGRFA	*International Treaty on Plant Genetic Resources for Food and Agriculture*
ITRC	Industrial Toxicology Research Centre (India)
IUPGRFA	*International Undertaking on Plant Genetic Resources for Food and Agriculture*
KCC	Kisan (Agriculturalist) Coordination Committee (India)
KRRS	Karnataka Rajya Raitha Sangha (India)
LMO	living modified organism
MMBL	Mahyco-Monsanto Biotech Limited (India)

MST	Movimento dos Trabalhadores Rurais sem Terra (Landless Rural Workers' Movement, Brazil)
NAFTA	*North American Free Trade Agreement*
NBAC	National Biotechnology Advisory Committee
NFU	National Farmers Union (Canada)
NGO	nongovernmental organization
OECD	Organisation for Economic Co-operation and Development
PBRA	*Plant Breeders' Rights Act*
PBRs	plant breeders' rights
PIPRA	Public Intellectual Property Resource for Agriculture
PNT	Plants with Novel Traits
PT	Partido dos Trabalhadores (Workers' Party, Brazil)
PTTC	Platform for Translational Research on Transgenic Crops
R&D	research and development
RCGM	Review Committee on Genetic Manipulation (India)
rDNA	recombinant DNA
RFSTE	Research Foundation for Science, Technology and Ecology (India)
RR	Roundup Ready (glyphosate-tolerant)
RRI	Responsible Research and Innovation
SMTA	Standard Material Transfer Agreement
TPP	*Trans-Pacific Partnership*
TRIPS	Trade-Related Aspects of Intellectual Property Rights
UPOV	International Union for the Protection of New Varieties of Plants
WEMA	Water Efficient Maize for Africa
WFD	Water Framework Directive (European Union)
WIPO	World Intellectual Property Organization
WTO	World Trade Organization

Acknowledgments

The research underlying this volume was completed as the "Genomics-related Ethical, Environmental, Economic, Legal and Social" (GE³LS) research portion of the Sunflower of Genomics project, and was funded by Genome Canada and Genome British Columbia. The editors want to thank our funders, as well as Loren Rieseberg, PhD, for his support of the GE³LS research. In addition, thanks are due to our long-time collaborator and friend, Ed Levy, PhD, for his always-stimulating contributions; to former research assistants Matthew Voell, JD; Rebecca Goulding, PhD; and to Regiane Garcia, JD, for moving the research forward; and to Milind Kandlikar, PhD, who graciously hosted the workshop from which this volume stems, at the Liu Institute for Global Issues at the University of British Columbia.

THE INTELLECTUAL PROPERTY–REGULATORY COMPLEX

Introduction

EMILY MARDEN, R. NELSON GODFREY, AND RACHAEL MANION

In recent years, the lack of success in translating significantly funded research insights into commercial products has been the subject of much discussion among policy makers, researchers, analysts, and funders. This concern has been voiced across the spectrum of genomics research, where significant investments in research have yet to generate the rates of return – in terms of identifiable economic measures – that were initially expected. This volume addresses this issue specifically in the agricultural arena with respect to potential causes of difficulties in translating funded agricultural genomics research into socially beneficial products.

Stakeholders and academics have identified many challenges to translating agricultural innovation arising from genomics research into products for end-users. Broadly speaking, such challenges include, but are not limited to, the cost of sustaining genomics research for the duration of time it takes to translate an innovation into a useful product; an inability in some cases to access inputs that are necessary for development but are protected by intellectual property (IP) rights; complex biosafety regimes that make any required product approvals expensive and difficult to obtain; and uncertainty about the market acceptance of innovations in light of sustained opposition to genetically modified organisms (GMOs) in agriculture. All of these issues have been explored in the past. To date, however, little attention has been paid to the particular challenges facing innovations arising specifically out of agricultural genomics beyond GMOs. Even within that context, less attention has been paid to the overall significance of what we call the "Intellectual Property–Regulatory Complex" or the "IP–Regulatory Complex."

In our view, the IP–Regulatory Complex refers to the sum total of domestic and international intellectual property and biosafety regimes pertaining to agricultural genomics, which operate independently of, and, more often than not, in tension with one another. We believe that the IP–Regulatory Complex materially impacts outcomes of (and funding for) innovation in agricultural genomics in ways that cannot be fully appreciated when examining intellectual property or biosafety regimes in isolation. In our view, without a full appreciation of the IP–Regulatory Complex, it will be difficult to address the potential barriers to implementing innovations in agriculture. This volume is intended as a first step in that conversation.

The IP–Regulatory Complex takes into account both IP and biosafety regimes. Within this context, IP refers to legal mechanisms for protecting and incentivizing innovations in agricultural genomics, which coexist with counter-movements to the proliferation of global intellectual property regimes, in the form of sharing regimes and mechanisms. Biosafety, in turn, refers to the regulatory regimes developed to ensure the human or environmental health and safety of innovations introduced into domestic or international agriculture. IP and biosafety regimes are distinct legal entities and are thus treated as distinct subjects in academic spheres and policy discussions. In reality, however, intellectual property and biosafety regimes *collectively* impact incentives for and potential outputs of agricultural genomics research and investment. Ultimately, as discussed in the chapters in this volume, the net collective impact of the relevant IP and biosafety mechanisms may actually run counter to the policy intent of a single IP or biosafety regime or framework. Furthermore, the net effect of the IP–Regulatory Complex as a whole may not be taken into consideration when research is funded, with the result that when innovations emerge or are intended for global use, initial assumptions made in the context of IP or biosafety may prove to be inadequate. Ultimately, a lack of appreciation of the IP–Regulatory Complex may frustrate uptake of genomics research.

When we speak of the IP–Regulatory Complex in this volume, we refer to the full range of domestic and international IP or biosafety regimes that govern innovations arising from agricultural genomics. Thus, in terms of biosafety regimes, the IP–Regulatory Complex includes domestic approaches, such as Canada's Plants with Novel Traits (PNT) regime (discussed in more detail in Chapter 1) as well as process- or product-based approaches in other jurisdictions. Many of these biosafety regimes were shaped, at least in part, as a response to perceived issues raised in the context of agriculture by recombinant technologies, or GMOs. Because national regimes take different approaches to biosafety regulation, the resulting domestic regulatory frameworks can be inconsistent with one another, both conceptually and often, as discussed in more detail in Chapter 4, in their application. Further, these domestic regimes sit within an international framework shaped by the *Convention on Biological Diversity* (CBD), discussed further in Chapter 1, which specifically allows for widely divergent application of biosafety principles. Taken together, the variability among biosafety regimes can greatly complicate efforts to develop and bring an innovative product to market.

These biosafety regimes coexist with myriad IP regimes that also impact the translation of innovations arising from agricultural genomics, but the two regimes were not developed to complement each other. In fact, the existence of IP regimes applicable to plants and agriculture represents a relatively recent legal development. For many years, proprietary rights for agricultural innovations were generally regarded as unavailable because germplasm was presumed to be a communal resource to be freely shared.[1] Thus, for most of recent history, agricultural practices permitted – and even expected – the saving and replanting of seeds and their sale to other farmers.[2] It was not until the development and widespread use of hybridization techniques[3] in the early twentieth century and the growth of a more robust seed industry that proprietary protections and in-kind responses to such protections became available for agriculture (Kloppenburg 1988, 11).[4]

With respect to proprietary rights, the *International Convention for the Protection of New Varieties of Plants* (the UPOV Convention) was signed in 1961 and adopted by most large agriculture- and seed-producing nations. Under the UPOV Convention, member nations are required to adopt legislation that institutes proprietary protection for all plant "varieties"[5] that are new, distinct, uniform, and stable (Article 5[1]).[6] Many countries have developed plant variety protection regimes that largely reflect the UPOV Convention provisions, although there is some diversity among them in the version of the convention adopted, and hence in their associated breeder's rights and farmer's privileges. Although conventional patent regimes have generally not been interpreted to be applicable to plants, a few countries, including the United States (as discussed further in Chapter 1), have deemed plants to be patentable subject matter. In these countries, then, plant variety and plant patent protection are both available.

At the same time, there are distinct international regimes that enshrine practices of sharing of germplasm in agriculture. Toward this end, the CBD and the *International Treaty for Plant Genetic Resources for Food and Agriculture* (ITPGRFA), discussed in greater detail in Chapter 5, expressly recognize the importance of proprietary rights in fostering innovation, particularly national proprietary interests in domestic genetic resources. In some cases, these mechanisms mandate the sharing of innovations derived from agricultural resources in a manner that may work at cross-purposes with proprietary IP protections (Marden and Godfrey 2012). The source of germplasm used for an innovation in agriculture can determine whether and to what extent proprietary protections are available, or whether, conversely, the innovation must be shared in a manner consistent with the CBD and the ITPGRFA.

Ultimately, to translate an innovation arising from agricultural genomics into the commercial sphere, the researcher or developer must come to terms with both biosafety regimes and IP regimes in jurisdictions of interest. This is no simple task. Each of the many domestic and international biosafety and IP regimes was developed in a certain context with a particular purpose at a specific time, resulting in layers of obligation and effect that stakeholders in the agricultural arena must address.

This volume represents an effort to arrive at a baseline understanding of the IP–Regulatory Complex, in terms of both its origins in the byzantine map of biosafety and IP regimes and its effects on innovation in the agricultural genomics arena. The contributors identify and examine aspects of these multiple regimes within the IP–Regulatory Complex that may affect agricultural genomics research. They also discuss how intellectual property and regulatory regimes may be viewed cumulatively as elements of the IP–Regulatory Complex, and speculate on impacts of the IP–Regulatory Complex on innovative research in the genomics space and the distribution of genomics products; in some cases, GMOs are used as a point of reference as such innovations have been present in the market and in the innovation cycle for a sustained period. With this understanding of the IP–Regulatory Complex as the foundation, solutions are proposed to facilitate greater efficiency in meeting the aims of IP and regulatory regimes while enabling innovative research (Chapters 7 and 8).

Part 1 of this collection articulates the concept of the IP–Regulatory Complex and introduces the current challenges posed by it. In Chapter 1, Emily Marden, Robert Nelson Godfrey, and colleagues review selected IP and regulatory regimes that may impact development of agricultural genomics products in the context of a publicly funded sunflower genomics project. They also discuss the challenges that the Canadian and international IP and biosafety regimes pose to such a project, one goal of which is to develop a sunflower intended for use in the developing world as an improved source of both food (from the plant's oil) and fuel (from a woodier stalk). The contributors to Part 1 also discuss regulatory instruments in the context of genomics and issues surrounding the translation of research into useful products. Building on the discussion of the potentially complicating role of biosafety regulation, they provide perspectives on the relevance of regulatory regimes to the development of genomics products, drawing on experiences with genetically modified organisms in particular.

In Chapter 2, Sarah Hartley provides a historical description of the development of Canada's regulatory framework for agricultural biotechnology. She reveals how regulators privileged science-based risk assessment to the exclusion of assessment strategies and perspectives incorporating social and ethical issues in agricultural biotechnology research. She discusses the important role of ethical considerations when making policies about agricultural genomics technology. As genomics research leads to the development of new avenues in agricultural biotechnology, different social and ethical issues may arise and present new challenges to exclusively science-based risk assessment approaches. According to Hartley, where the federal regulatory framework fails to provide a venue for considering such challenges at an early stage, the robustness and legitimacy of the regulatory regime will be undermined, even as it strains to incorporate newly developed genomics techniques and technologies.

In Chapter 3, Gregory Graff and David Zilberman explore the incentives for innovation created by the IP–Regulatory Complex. They discuss how IP and regulatory systems mutually reinforce one another, and how they can extend the control of a few innovators and favour large markets. The analysis is informed by an empirical study of products of agricultural genomics research that meet a notional definition of "social benefit," such as crops having drought and other stress tolerance traits, crops with improved nutritional content, and crops for clean energy – all of which are areas in which the authors note promising recent endeavours. In order to better understand how to incentivize the development of these socially beneficial crops, the authors explore the impact of the IP–Regulatory Complex on the choices of which crops and traits get developed in the absence of the same profit incentives offered by major crops.

Importantly, as Ron Herring makes clear in Chapter 4, the existence of a regulatory regime per se may have limited impact on how products are adopted and used in any given context. He explores the powerful biopolitics surrounding the commercial introduction of genetic technologies in the developing economies of India and Brazil. He also discusses the impact of a wide array of stakeholders, events, and decisions on the development of local biosafety and bioproperty regimes in India and Brazil. In those countries, the desire of farmers for a cheap alternative to *Bt–* and Roundup Ready–modified crops led them to import, exchange, and use "stealth seeds," which were modified to contain the desired traits despite the regulatory and IP regimes enacted to prevent such activity, thereby undermining the capability and legitimacy of those regimes.

Chapter 4 provides a reality check to debates over policy making for socially beneficial biotechnology and genomics products. Rural farmers are often represented in the discussions that shape such policies, but Herring points out the disconnect between the poor farmer and the activist who claims to represent her. Furthermore, Herring's observations on the flourishing trade involving stealth seeds in India and Brazil raise important questions about the practicality and power of biosafety regimes on the ground in developing countries.

Similar to biosafety regulation, current IP mechanisms are particularly relevant to emerging research in genomics. The chapters in Part 2 provide important perspectives on intellectual property elements of the IP–Regulatory Complex: that of the innovative entity seeking to manage its intellectual property in a global marketplace, and that of the counter-movement toward seed sharing and the relevance of the ITPGRFA toward that end.

In Chapter 5, Chidi Oguamanam discusses the implementation experience of the ITPGRFA and associated regimes. He reviews the treaty's major contributions to the existing policy landscape in areas including biosafety, intellectual property, and access and benefit sharing. He takes the position that the treaty and related regimes help to moderate or balance the generally restrictive effects of intellectual property regimes; provide an as yet untested opportunity for greater access to and benefit sharing of innovations and profits in emerging fields, including agricultural genomics; and incentivize research for socially beneficial purposes.

In Chapter 6, Jeremy Hall, Stelvia Matos, and Vernon Bachor analyze the commercialization of genetic technologies in India and Brazil, in the process critiquing intellectual property protection strategies employed by biotechnology firms in those countries. They suggest that the firms that led the commercialization of transgenic technology in Brazil and India employed IP management strategies that failed to adequately consider both cognitive and socio-political legitimization processes. They also suggest that the institutional policy framework surrounding agricultural biotechnology is complex, with numerous stakeholders representing varied interests throughout. Applying the experience of biotechnology firms to the emerging field of genomics research and commercialization, Hall and colleagues conclude that early-stage IP management strategies – considering not only economic viability and commercial feasibility but also issues relating to social legitimacy and institutional differences – are necessary to assist in the diffusion and commercial adoption of next-stage technologies.

In light of the complicated interactions between composite biosafety and intellectual property regimes observed in Parts 1 and 2, the chapters in Part 3 recommend strategies for overcoming some aspects of the IP–Regulatory Complex. The diverse nature of the proposed solutions and the target audiences of the chapters reflect the intricate entanglements of the international and domestic IP and biosafety regimes that comprise the IP–Regulatory Complex.

Building on many of the democratic deficits noted by Sarah Hartley in Chapter 2, Regiane Garcia discusses the limitations of the current approach to technical and scientific regulation in Chapter 7. She describes the policy decisions in the current approach as made by experts to the exclusion of the concerns of many stakeholders and citizens, and raises the question of whether regulatory approaches to biotechnology and genomics should be fundamentally retooled in order to be more democratic and inclusive. Drawing on her background in public health policy and governance, Garcia contrasts the classic "regulatory" model with a "new governance" approach to regulatory policy making, and recommends the European Union's *Water Framework Directive* as a possible model for genomics policy that is more in line with the "new governance" approach to policy making.

Finally, in Chapter 8, Rochelle Dreyfuss examines the discordant elements of domestic and international IP regimes as they apply to products of agricultural genomics research. She notes the difficulties in balancing interests in a domestic IP regime, a problem that is compounded in the international arena, where international IP agreements are often conceptualized and concluded in the context of trade discussions. Building on her work with Graham Dinwoodie titled *A Neofederalist Vision of TRIPS: The Resilience of the International Intellectual Property Regime,* she reviews the core principles that emerge when the IP system, as it applies to agricultural genomics research, is viewed holistically. She suggests that such an approach could preserve national flexibilities in IP law and contribute to a more internally harmonized IP system, where policy makers and decision makers have a more informed understanding of the balance struck between different interests and the flexibilities available in making policy. Dreyfuss suggests that flexibilities in national IP regimes could, in turn, provide mechanisms to balance the impacts of biosafety regimes on products of agricultural genomics research.

Taken together, these chapters present domestic and international regulatory and IP regimes as intimately intertwined. Without a deeper appreciation of the impact of the IP–Regulatory Complex on the development of agricultural genomics products, reforms to each regime alone may not achieve the intended social benefits, and the promise of these technologies may not be fully realized.

Notes

1 Originally the patent system was viewed as an inappropriate restriction on plant life (Llewelyn 1997). The historical role of intellectual property in agriculture was its application to mechanical inventions. Protection of plant varieties developed only recently, and was shaped by the prevailing economic conditions of developed countries (Commission on Intellectual Property Rights 2002). See also Pray and Naseem 2003, which describes the shift from public investment (for the public good) in agricultural research to private (for-profit) investment, and its direct impact on how developing countries access agricultural technology, especially seeds, which they have traditionally accessed through sharing or saving methods.

2 The historical practice of seed saving was a fundamental tenet of agriculture with far-reaching cultural significance (Belsie 1998); see also Weiss 1999 for a description of the role of the United States Department of Agriculture, land grant colleges, and extension services in development of new seed varieties. This research was publicly financed and patents were seldom sought or enforced.

3 "Hybridization, or scientifically combining and breeding seeds, was the first method by which companies were able to control replanting of seeds. For the first time, farmers were able to purchase improved seeds for a better crop. The drawback was that the second generation of crops did not fare as well as the first generation" (Stein 2005).

4 Stein (2005, 164–68) traces the private industry and judicial influences in the development of modern proprietary protections.

5 "Variety" is defined as "a plant grouping within a single botanical taxon of the lowest known rank, which grouping ... can be defined by the expression of the characteristics resulting from a given genotype or combination of genotypes, distinguished from any other plant grouping ... and considered as a unit with regard to its suitability for

being propagated unchanged" (UPOV Convention, Article 1[vi]). It is worth noting that rights under the UPOV Convention extend to all novel varieties that meet the criteria, regardless of how they are derived, and are thus relevant for varieties characterized by or developed through genomics techniques and/or biotechnology (see Article 5).

6 To be "new," a plant variety cannot have been offered for sale or marketed earlier than one year before the application for protection is filed in the source country, or for a period longer than four years in any other country (UPOV Convention, Article 6[1]). To be "distinct," the variety must be "distinguishable from any other variety whose existence is a matter of common knowledge" anywhere in the world (Article 7). Varieties that are "common knowledge" are defined rather simply as any plants that meet the definition of "variety" in Article 1(vi), and includes varieties that have not obtained protection according to the UPOV Convention (International Union for the Protection of New Varieties of Plants 2002). To be "uniform" and "stable," the variety's relevant characteristics must remain true and sufficiently uniform on repeated propagation, subject to the variation that may be expected due to the particular features of its propagation (Articles 8 and 9). See also Dutfield 2008, 35, explaining that "the uniformity requirement also shows the specific nature of the UPOV system, since this requirement cannot practically be the same for species with different ways of reproduction."

References

Treaties and Agreements
International Convention for the Protection of New Varieties of Plants, December 2, 1961, 815 U.N.T.S. 89 (as revised at Geneva on November 10, 1972, on October 23, 1978, and on March 19, 1991) [UPOV Convention].
International Treaty on Plant Genetic Resources for Food and Agriculture, adopted November 3, 2001. S. Treaty Doc. No. 110-19 [ITPGRFA]. ftp://ftp.fao.org/docrep/fao/007/ae040e/ae040e00.pdf.

Secondary Sources
Belsie, Laurent. 1998. "Plants without Seeds Challenge Historic Farming Practices." *Christian Science Monitor,* July 30, 1998. http://www.csmonitor.com/1998/0730/073098.feat.feat.6.html.
Commission on Intellectual Property Rights. 2002. *Integrating Intellectual Property Rights and Development Policy.* London: Commission on Intellectual Property Rights. http://www.iprcommission.org/papers/pdfs/final_report/CIPRfullfinal.pdf.
Dutfield, Graham. 2008. "Turning Plant Varieties into Intellectual Property: The UPOV Convention." In *The Future Control of Food: A Guide to International Negotiations and Rules on Intellectual Property, Biodiversity and Food Security,* ed. Geoff Tansey and Tasmin Rajotte, 27–47. London: Earthscan.
International Union for the Protection of New Varieties of Plants. 2002. *The Notion of Breeder and Common Knowledge,* C(Extr.)/19/2 Rev., Annex, paras. 22–24 (August 9, 2002).

Kloppenburg, Jack Ralph Jr. 1988. *First the Seed: The Political Economy of Plant Biotechnology, 1492–2000.* New York: Cambridge University Press.

Llewelyn, Margaret. 1997. "The Legal Protection of Biotechnological Inventions: An Alternative Approach." *European Intellectual Property Review* 19 (3): 115–27.

Marden, Emily, and R. Nelson Godfrey. 2012. "Intellectual Property and Sharing Regimes in Agricultural Genomics: Finding the Right Balance for Innovation." *Drake Journal of Agricultural Law* 17 (2): 369–93.

Pray, C.E., and A. Naseem. 2003. *The Economics of Agricultural Biotechnology Research.* ESA Working Paper No. 03–07. Rome: Food and Agriculture Organization of the United Nations.

Stein, Haley. 2005. "Intellectual Property and Genetically Modified Seeds: The United States, Trade, and the Developing World." *Northwestern Journal of Technology and Intellectual Property* 3 (2): 160–78.

Weiss, Rick. 1999. "Seeds of Discord; Monsanto's Gene Police Raise Alarm on Farmers' Rights, Rural Tradition." *Washington Post,* February 3, 1999. http://www.artsci.wustl.edu/~anthro/articles/monsanto1.html.

PART 1

PERSPECTIVES ON REGULATORY REGIMES

1

Biosafety and Intellectual Property Regimes as Elements of the IP–Regulatory Complex

The Case of Canadian Sunflower Genomics

EMILY MARDEN, R. NELSON GODFREY,
MATTHEW R. VOELL, AND LOREN H. RIESEBERG

In recent years, governments have made significant investments in genomics research on agricultural crops and species. For example, Genome Canada, a Canadian funding body instituted to develop and support Canadian-based proteomics and genomics research, has funded a number of large-scale projects to sequence the genomes of a wide array of species (Genome British Columbia 2013). The United States government has likewise funded genomics research programs on soy, corn, cotton, and wheat, among others, and government grants supporting plant genomics research have reached an all-time high (Waltz 2010, 10; National Science Foundation 2010). According to public researchers, industry leaders, and funders alike, the information resulting from these large-scale projects has the potential to jump-start innovation in agricultural genomics as well as accelerate developments in conventional breeding. Genomics data can lay a foundation of information useful for genetic engineering, but are not necessarily tied to it. Indeed, an understanding of a plant's genome can help scientists better understand the location of traits and thus enable more fruitful conventional breeding to yield desired traits. This research may generate insights that provide wider benefits both domestically and in developing countries.

Despite this significant public investment in genomics research on agricultural crops and species, relatively little attention has been paid to how existing legal frameworks affect the realization of national research goals. In particular, there has been minimal consideration of the collective role of

applicable biosafety and intellectual property (IP) regimes on product development in the agricultural genomics arena.

In general, biosafety regulations are directed toward ensuring the safe
development and market entry of novel plant products. Intellectual property regimes have a different goal – that of creating property interests in
innovative products in a manner that provides incentives for development
and encourages access to these innovations by users. Yet, developers of new
products must consider navigating both biosafety regimes *and* IP regimes
while developing a strategy to make their innovations widely available (commercially or otherwise). Thus, to the extent that biosafety regulations add a
complex set of approval obligations, IP regimes may be necessary to max-
imally capture value to deliver a return on investment. In an analogous
fashion, it may be that where biosafety issues are minimal, a developer
may opt to facilitate uptake through less proprietary approaches to IP.
Throughout this volume, the net operation of biosafety and IP regimes *in
toto* on a particular product (both national and international) is referred to
as the "Intellectual Property–Regulatory Complex" or the "IP–Regulatory
Complex."

In general, the academic literature has paid little attention to the impact of the Intellectual Property–Regulatory Complex on publicly funded
agricultural genomics, particularly with respect to publicly funded projects
that, unlike projects in industry, may lack the potential for significant economic return (whether in terms of domestic market return or because their
products are intended primarily for export to developing countries). Moreover, where domestic or international regulatory or IP regimes have been
analyzed, they have generally been analyzed separately rather than as con-
stituent elements of an "IP–Regulatory Complex" that affects the development and delivery of such products.[1]

In this chapter, we provide an overview of selected international IP and
regulatory regimes that may be applicable to the development of agricultural products arising from (largely publicly funded) genomics research. We
then narrow the focus to similarly applicable IP and regulatory regimes in a
particular nation – Canada – to illustrate how the complex of these regimes
operates at the national level through the implementation of national policies corresponding to the international regimes. We draw on our own experience as the social science research team for a sunflower genomics project,
where we have found that a putative dual-use food/fuel sunflower would
face potentially insurmountable obstacles due to multiple layers of biosafety
and IP regimes. The variety of national schemes that have been designed to

implement the key international instruments surveyed here is another element of the IP–Regulatory Complex that can complicate product development and increase upstream costs. A number of contributors to this volume have framed their analyses and case studies around Canadian policy choices, but other national approaches to IP and regulation of genomics and related products, and their relevance to the notional "IP–Regulatory Complex," are discussed where appropriate throughout this volume.

In the next section, we explore the nature of the sunflower, its importance as an agricultural/horticultural crop, and its potential as a socially beneficial dual-use food/fuel crop. Next, we introduce certain international legal instruments, comprising international regulatory and trade instruments, on the one hand, and intellectual property and related regimes, on the other. These regimes include a suite of World Trade Organization (WTO) agreements (foremost among them the *General Agreement on Tariffs and Trade* [GATT] and the *Agreement on the Application of Sanitary and Phytosanitary Measures* [SPS Agreement]); the *Convention on Biological Diversity* (CBD) and the *Cartagena Protocol on Biosafety* (Secretariat of the CBD 2000); and, in the context of intellectual property, the *International Convention for the Protection of New Varieties of Plants* (UPOV Convention), the *Agreement on Trade-Related Aspects of Intellectual Property Rights* (TRIPS Agreement), and the *International Treaty on Plant Genetic Resources for Food and Agriculture* (ITPGRFA).

Following this, we review Canadian IP and regulatory regimes that may impact the approval of and protections available for agricultural products, as a case study of national policy making in the context of the foregoing regimes. These Canadian regimes include the following: in the biosafety context, the Plants with Novel Traits (PNT) and Novel Foods regimes under the *Seeds Act* and *Food and Drugs Act* and their associated regulations; and, in the context of IP protection, the *Plant Breeders' Rights Act* and case law pointing to the patentability of certain plant matter under the *Patent Act*. Finally, we offer some suggestions on the net effects of this multilayered legal landscape as a "complex" affecting the development and delivery of products arising from agricultural genomics research, particularly as it pertains to the development of the putative dual-use sunflower genomics product discussed earlier in the chapter.

Background

The common sunflower (*Helianthus annuus* L.) is a globally important oilseed, food, and ornamental crop. The species ranks eleventh among world crops in terms of total acreage, with a farmgate value between US$7 billion and US$11 billion (Food and Agriculture Organization 2014).[2] Sunflower oil is considered to be a premium vegetable oil because of its high unsaturated fatty acid composition, high levels of vitamin E, and low linolenic acid content.

That the sunflower could also be a source of firewood or charcoal has only recently been recognized. The closest relative of the cultivated sunflower is a drought-tolerant annual species, the silverleaf sunflower (*Helianthus argophyllus*), known for its woody stalks that grow three to four metres tall and up to ten centimetres in diameter in a single season. Domesticated sunflower varieties with this trait would enable farmers to grow small, woody "trees" in a single year. The sunflower seeds could still be harvested for food, but the woody stalks could also be used for fuel. The wood is similar in quality to the wood of quaking aspen, suggesting that it could also serve well as building material or roof thatching.

The Genomics of Sunflower project, funded by Genome Canada and Genome British Columbia, is an international research program that aims in part to understand the genetics of wood formation in sunflowers and to develop woody sunflower cultivars that are well adapted to the conditions of subsistence agriculture. The wood-producing cultivars will be developed using traditional sexual breeding approaches, which may be accelerated through marker-assisted selection. Marker-assisted selection can also ensure that other chromosomal segments, which may have negative effects on seed or oil quality, are not accidentally introduced during the breeding process. The research team aims to direct germplasm from the most productive woody cultivars to several agricultural research institutes across sub-Saharan Africa for evaluation and distribution.

Although genetic analyses are well underway, no studies have been performed to determine whether there is a trade-off between wood production and yield or whether wood production would make wild sunflowers more weedy or invasive. It seems likely, however, that the diversion of energy from seed to stem development would lead to a reduction of oil yield in cultivated plants, as well as to reduced reproductive fitness in wild or weedy populations. Thus, while rates of hybridization between cultivated and wild sunflowers can be as high as 37 percent (Arias and Rieseberg 1994), and the fitness disadvantage of F1 hybrids is fairly small (Snow et al. 1998), researchers believe that wood-producing genes are unlikely to escape and spread because of their negative effects on fitness.

International Regulatory and Intellectual Property Regimes

The regulatory pathways required for approval of novel agricultural products are designed to achieve multiple, often conflicting ends. Thus, for example, Canadian regulation of novel agricultural products reflects both an attempt to minimize regulatory burdens on the agricultural sector (see Abergel 2007 for a summary emphasizing the importance of the agricultural industry to the Canadian economy and the necessary export market accessibility as reasons for the deregulatory position of the Canadian government in the agricultural sector), and an effort to ensure the safety of products for humans and the environment. Alternatively, some international regulatory regimes represent efforts to ensure that free trade is not impeded by unfounded barriers to new agricultural products, whereas coexisting international regimes allow the same governments to place import or development restrictions around novel technologies in the interests of national environmental and health biosafety.

International Regulatory Framework

In the international policy arena governing the development of agricultural products, the relevant regimes work at cross-purposes to some extent. With respect to trade, international and bilateral agreements (including the *General Agreement on Tariffs and Trade* and the *Agreement on the Application of Sanitary and Phytosanitary Measures*) play a significant role in national biosafety policy making. At the same time, the *Convention on Biological Diversity* outlines biosafety and health measures for novel plants. These different agreements coexist and coordinately affect policy development despite the inherent tensions between them. We begin with a brief examination of the trade regimes developed by the World Trade Organization that apply to its global membership base.

The GATT impacts any commodity that moves in international trade, including seeds and products arising from agricultural research. It promotes a liberal trade environment, which may be restricted only by individual member states' national policies in limited circumstances. To that end, the GATT permits national regulatory measures beyond taxes, duties, and tariffs (Article XI) only where such measures are *necessary* to protect human, animal, or plant health (Article XX[b], emphasis added).

The SPS Agreement expounds on the GATT exemption, requiring that "any sanitary or phytosanitary measure [regulation] is applied *only to the extent necessary* to protect human, animal or plant life or health" (Article 2.2, emphasis added) and that "such measures [regulations] are *not more trade-restrictive than required* to achieve their appropriate level of sanitary or phytosanitary protection" (Article 5.6, emphasis added).[3] A trade measure is "more trade restrictive than required" if there is another measure that is reasonably available, taking into account technical and economic feasibility, that achieves the appropriate level of protection and is significantly less restrictive to trade (Article 5.6, footnote 3). Further, any measures must be supported by "scientific justification" and may not be inconsistent with any other provision of the SPS Agreement (Article 3.3, footnote 2). Such measures may not be inconsistent with Article 5.1 or other substantive provisions of the SPS Agreement (*Appellate Body Report, EC Measures Concerning Meat and Meat Products,* para. 186). The GATT and other WTO agreements thus stand as legal regimes that, for the most part, prevent member countries from regulating trade in agricultural products except where demonstrably necessary for health reasons.

The CBD, in contrast, aims to generally reduce or eliminate risks posed to a nation's biological diversity by particular products and categories of activities. Thus, Articles 7 and 8 mandate that parties "identify processes and categories of activities which have or are likely to have significant adverse impacts on the conservation and sustainable use of biological diversity" (Article 7[c]), and "regulate or manage the relevant processes and categories of activities" (Article 8[1]). To those ends, Article 14(a) requires that parties "introduce appropriate procedures requiring environmental impact assessment of its proposed projects that are likely to have significant adverse effects on biological diversity." As a result of the CBD, numerous countries, particularly those rich in biodiversity, have complex regulations pertaining to how plant material and genetic resources are imported, exported, and shared between nations (Carrizosa et al. 2004).

It is worth noting that although the CBD applies in some measure to all novel plants, regardless of how they are developed, specific provisions exist for agricultural biotechnology products. The CBD requires that parties "establish or maintain [the] means to regulate, manage or control the risks associated with the use and release of living modified organisms resulting from biotechnology" (Article 8[g]). The *Cartagena Protocol on Biosafety,* an addition to the CBD, allows for further regulation with respect to genetically modified plants – referred to in the protocol as "living modified organisms" (LMOs)[4] – that "*may* have adverse effects on the conservation and sustainable use of biological diversity, taking also into account risks to human health" (Article 4, emphasis added; see Secretariat of the CBD 2000).

Controversially, the protocol goes on to permit precautionary measures for regulating LMOs as long as they are conducted in a scientifically sound manner and take into account recognized risk assessment techniques.[5]

The net impact of international biosafety and trade regimes on the development and delivery of genomics products is difficult to address summarily. Nations emphasize trade and biosafety priorities to differing degrees, in line with national trade priorities, policy goals, social preferences, and environmental conditions. Some nations agree with the WTO emphasis on free markets, whereas others draw on the precautionary measures permitted by the *Cartagena Protocol* and the CBD (Safrin 2002; Phillips and Kerr 2000). The net impact of the sometimes overlapping and often distinctive biosafety and trade regimes is to present developers of genomics products with a daunting and complicated portion of the IP–Regulatory Complex to be considered in concert with domestic biosafety legislation and intellectual property regimes. We return to the question of domestic biosafety regimes later, in our discussion of Canada's biosafety policies.

International Intellectual Property Regimes

Regulatory regimes are directed toward ensuring the safe development and market entry of novel plant products. Intellectual property regimes, by contrast, have the goal of creating incentives for innovation in exchange for the public disclosure of new technological developments. Thus, IP regimes play a critical role in facilitating, and creating incentives for, the development of products from agricultural genomics. It is important to recognize that national decisions on the form and substance of IP regimes can have significant downstream impacts on the uptake and ultimate utility of innovations (see, e.g., Rimmer 2008; Van Overwalle 2009; Hope 2008; Goulding et al. 2010). In this section, we briefly outline the possible forms of IP protection for plants available internationally. Later in this chapter, we review the IP laws applicable to the development of novel plant varieties in Canada, as a case study of national implementation of the international agreements surveyed here.

The application of IP regimes to plants is a relatively recent development. Historically, there was a general consensus that it was better to maintain the free exchange of new plant materials and information in order to ensure the widest dissemination of the best possible plant varieties (Llewelyn 1997, 117; Pray and Naseem 2003). It was not until the mid-twentieth century, and the emergence of a robust seed industry seeking to create property rights in its innovations, that intellectual property protection for plants became a subject of policy making.

There are several aspects to the application of IP regimes to plants that emerged thereafter. The first aspect stems from early efforts by plant breeders to attain some form of recognition; this came about in so-called Plant Breeders' Rights enactments. The second arises from the expansion of patent law in certain jurisdictions to include plant patents, particularly over the last several decades, and the third relates to a corresponding increase in the number of regimes dedicated to sharing of plant IP and genetic resources. These three aspects (and their corresponding regimes) stand in an awkward balance, simultaneously multiplying conceptual approaches to IP and complicating incentives for conducting innovative research and early-stage investments, including early-stage expenses incurred in overcoming biosafety hurdles.

The UPOV Convention

In an effort to capitalize on increased market possibilities, seed companies began pressing policy makers in the mid-twentieth century to develop a legal regime for protecting IP rights in plants. The resulting early form of IP protection for plants was intended to reward breeders for developing and disclosing new plant varieties.[6] After much debate and discussion, these so-called plant breeders' rights were formalized in the *International Convention for the Protection of New Varieties of Plants* (known as the UPOV Convention), which was agreed on by a group of European seed-producing countries in Paris in 1961 (revised in 1978 and 1991) and entered into force in 1968 on receiving the requisite number of signatures.

Intellectual property rights under the UPOV Convention are defined as "breeders' rights." Parties to the UPOV Convention are required to introduce legislation that grants breeders the exclusive rights to produce or reproduce, offer for sale, or export protected varieties (Article 14[1]). To be eligible for IP protection under the UPOV Convention, plant varieties must be new, distinct, stable, and uniform. To be new, a plant variety cannot have been offered for sale or marketed in the source country more than one year prior to the filing of an application for a breeder's right, or marketed for a period longer than four years prior to the filing of an application for a breeder's right in any other country. To be distinct, the variety must be distinguishable from any other variety known anywhere in the world. Further, the variety must remain uniform and stable, that is, true to its description after repeated reproduction or propagation.

Other key parameters built into the UPOV Convention include: (1) a breeders' exemption – present in both the 1978 and 1991 revisions – which effectuates a research exemption under which plant breeders may access protected material for the purpose of developing new varieties; and (2) a farmers' privilege (also commonly referred to as "farmers' rights"), which allows farmers to save and replant seed.[7] Farmers' rights, in the context of international accords such as the UPOV Convention, are discussed in greater detail in Chapter 5. The UPOV Convention has become a model for plant variety protection regimes in some countries, and was more widely adopted as a generic plant protection system subsequent to the passage of the TRIPS Agreement.

The TRIPS Agreement

Even as plant breeders' rights regimes developed, some viewed the patent system as inappropriate to protect new plant varieties, as it seemed unlikely that innovators could meet the novelty, inventive step, disclosure, and, perhaps most importantly, the subject matter requirements for patent protection. More recently, however, the scope of IP protection for plants has expanded in a number of jurisdictions concomitant with the expansion of patent rights to life forms in general.[8]

The emergence of the *Agreement on Trade-Related Aspects of Intellectual Property Rights* as a force in plant IP flows from the same historical current. In essence, the TRIPS Agreement sets out the substantive minimum standards for patent protection writ large for all WTO members, and provides mechanisms for enforcement and dispute resolution. While the TRIPS Agreement provides in Article 27(1) that patents shall be available for all products and processes without discrimination as to the field of technology, Article 27.3(b) allows members to exclude from patentability "plants and animals other than micro-organisms, and essentially biological processes for the production of plants or animals other than non-biological and microbiological processes." Most important for these purposes, however, Article 27.3(b) also states that "members shall provide for the protection of plant varieties either by patents or by an effective sui generis system *or any combination [thereof]*" (emphasis added).

Article 27.3(b) has been criticized by some as the "biotechnology clause," which some argue reflects the interests of developed countries in ensuring protection for innovations related to life forms (and biotechnology) in response to concerns that developing countries may exclude such life forms from patentability (Roffe 2008, 59–60). To date, the scope of Article 27.3(b) has not been definitively articulated and thus there is little clarity on what constitutes an "effective" plant variety protection system. The net result is that, globally, there exists an assortment of different and often inconsistent

regimes for IP protection of plants, which may make it complex to plan for
the development and protection of plant varieties except on a country-
by-country basis (see, for example, Thorpe 2002 on the adoption of the
UPOV Convention or other *sui generis* systems, and the existence of incon-
sistencies in the implementation of farmers' rights policies). Further, even if
a plant were subject to IP protection in a country such as Canada, there can
be no guarantee that the same scope of protection would be available in
other importing countries.[9]

To some extent, countries have endeavoured to minimize these inconsis-
tencies by negotiating bilateral and regional free trade agreements that address uncertainties in general and the patenting of life forms in particular
(Rajotte 2008, 142). For example, some free trade agreements between the
United States, the European Union, and developing countries contain provisions that extend IP protection beyond the minimum obligations set out
in the TRIPS Agreement (Rajotte 2008). Such agreements are often characterized as "TRIPS-plus agreements," as many of them require parties (most
often developing countries) to introduce patent protection for plants and
animals (Rajotte 2008, 142) or, at the very least, to adopt the 1991 version of
the UPOV Convention. In this sense, TRIPS-plus agreements may override
IP choices mutually agreed on in the international arena and add another
element in evaluating the IP–Regulatory Complex as it operates in particular nations (Chapter 8; Dutfield 2008, 38–39).

International Treaty on Plant Genetic Resources for Food and Agriculture

A third major approach to plant IP represents a counter-movement to the
rapid proliferation of plant property rights, and is geared instead toward
facilitating the exchange and open availability of innovations and products
(Oguamanam 2006). Thus, for example, mechanisms guaranteeing access
to "plant genetic resources" have emerged in response to the proliferation
of the plant IP rights precipitated by the UPOV Convention and the TRIPS
Agreement. Here we discuss the International Treaty on Plant Genetic
Resources for Food and Agriculture (ITPGRFA) as representative of such
regimes, although numerous other agreements also exist.[10]

The ITPGRFA was developed, *inter alia,* as part of an effort to ensure that
sui generis obligations under TRIPS – in addition to emerging patent rights
in countries such as the United States – do not have the undesirable effect of
impeding access to plant and agricultural genetic resources essential for ongoing innovation (Marden and Godfrey 2012). To this end, the treaty creates

a "genetic resource commons," under which signatories have agreed to make "listed materials" (including sunflower resources and most major food crops) freely available to other signing countries "solely for the purpose of utilization and conservation for research, breeding and training for food and agriculture" (Article 12.3[a]). These obligations are limited to resources that are "under the management and control of the Contracting Parties and in the public domain" (Article 11.2), and thus do not affect privately held resources.

With respect to IP rights, the treaty states that "recipients shall not claim any IP or other rights that limit the facilitated access to the PGRFA [plant genetic resources for food or agriculture] or their genetic parts or components, in the form received from the multilateral system" (Article 12.3[d] and the ITPGRFA Standard Material Transfer Agreement [SMTA]). Anecdotal evidence suggests that parties in developed countries, such as the United States and Canada, interpret this provision as broadly permitting patenting of improvements and derivatives in any form except that received from the multilateral system. Others, including developing countries, may see it as not permitting any IP protection on derivatives or improvements by dint of the "genetic parts or components" language. There is currently no clear and agreed-on interpretation (Halewood and Nnadozie 2008, 122).

Because the treaty is a relatively new international mechanism, there is little precedent for its use as a mechanism for dissemination of genetic resources rather than simply as an open repository. The commons created by the treaty, however, is being expanded on significantly by the efforts of Bioversity International, a research-for-development organization dedicated to sustaining biodiversity, and is already being drawn on by developers (Oguamanam 2006). At present, the treaty may serve as an alternative to IP regimes as a means of ensuring open access to and distribution of some important food and feed crops in some circumstances. More broadly, it exemplifies the international movement in favour of guaranteeing the free availability of essential plant genetic resources unencumbered by proprietary rights. A further perspective on the impact of the treaty on the international regulatory and IP landscape and the IP–Regulatory Complex in general is found in Chapter 5 of this volume.

We turn now to a case study of national implementation of the international regimes outlined above, in the form of Canada. We stress that the Canadian policies briefly examined here are but one example of such national implementation, and should be viewed as an interesting example of how the IP–Regulatory Complex plays out on the national level rather than as any sort of guide to the kinds of policies that may apply in other nations.

Canadian Intellectual Property and Regulatory Regimes

To confuse matters further for developers, states generally do not adopt international regimes wholesale or otherwise interlock easily with international regulatory or trade regimes, including those surveyed above. Nor do states generally adopt policies that are consistent with those of their neighbours, basing their policy choices on national priorities instead. The net result is that national regulatory measures add a further layer of complication and difficulty to the complex created by international regulatory regimes. Nationally as well as internationally, regulatory regimes are complicated, often in tension with their proffered aims, and ultimately burdensome to any party attempting to develop novel agricultural products. This is certainly the case with respect to putative products developed from publicly funded agricultural genomics research.

Canada is party to the *Convention on Biological Diversity* and the WTO, and has enacted regulatory regimes that, in their own way, implement Canada's international regulatory and trade obligations. Although numerous Canadian statutes may affect the development of agricultural products intended for different purposes, such as those intended for consumption as food, all novel plants developed within Canada, regardless of the intended purpose of the products and whether or not such products are intended to be used in Canada, are subject to the "Plants with Novel Traits" (PNT) regulatory regime.[11] The PNT regime will form the bulk of our analysis of domestic Canadian regulatory provisions.

Canadian Regulation of Plants with Novel Traits

The Canadian Food Inspection Agency (CFIA) is the primary Canadian regulatory authority tasked with addressing novel agricultural products, including those arising from genomics. Its mandate includes preventing damage to the environment and safeguarding human health (CFIA 2012a). The CFIA acts within the authority delegated to it by the *Seeds Act*. "Seed" is defined in section 2 of the act as "any part of any species belonging to the plant kingdom, represented, sold, or used to grow a plant"; for the purposes of this chapter, the words "plant" or "crop" will be used to represent any part of a potential PNT's life cycle. Under the authority of the *Seeds Act*, the CFIA's regulations apply a *product*-based approach, under which regulators evaluate the end products of innovation rather than issues raised specifically by the *process* with which they were developed. This approach was developed in conjunction with a similar approach in the United States in an explicit effort by both countries to maintain leadership in the development of new

agricultural traits and to ensure that such products are treated in a consistent manner regardless of their means of development (Marden 2003).

Within this framework, Canada's particular regulatory approach casts
the "novelty" of a plant as the trigger for regulatory review. PNTs are defined
as any plants "containing a trait not present in plants of the same species
already existing as stable, cultivated populations in Canada, or present at a
level significantly outside the range of that trait in stable, cultivated populations of that plant species in Canada" (CFIA 2004, s. 1; *Seeds Regulations*,
s. 107). Thus, while the regulations are intended in some measure to not
unduly burden innovative approaches to plant development, the breadth of
techniques and products subject to them is large given the ambiguity in and
scope of the definition and the range of potential new traits possible (Smyth
and McHughen 2008, 223).

In Canada, the "novelty" of any new variety is identified and assessed by
the developer of the plant, with the requirement that all relevant information
be taken into account (CFIA 2009, ss. 2.1–2.2).[12] Plants that are not considered novel, and therefore need not be approved or reviewed by the Plant
Biosafety Office (PBO), are those that are considered *both* familiar to the
Canadian regulators and substantially equivalent to seed of the same species
(*Seeds Regulations*, s. 108[c]; Barrett and Abergel 2000). The concepts of
"familiarity" and "substantial equivalence" are distinct and considered indi-
vidually.[13] If a plant's characteristics are unfamiliar to Canadian regulators,
the plant is novel and must undergo a safety assessment by the PBO before
being released (CFIA 2009). If the relevant analogue is known to Canadian
regulators, any plant that is not substantially equivalent to that known quantity (in terms of its potential weediness, potential to develop into a plant pest,
potential to cause gene flow to native organisms, and potential to otherwise
negatively affect nontarget organisms or biodiversity generally; CFIA 2009,
s. 2.3.2) is novel and must undergo a safety assessment.[14] To complete its
safety assessment, the PBO compares data submitted by the developer with
species-specific companion documents that approximate scientific consensus on the baseline characteristics of the species. The criteria applied
by the PBO in the course of the assessment are the same as those involved
in determining novelty – degree of familiarity and substantial equivalence
to familiar species – and the PBO applies principles of risk assessment
scientifically, consulting relevant experts when necessary (CFIA 2004, s. 6).

It is worth noting that views on the PNT regulatory approach vary widely.
Some claim that the novelty-based approach makes the regulatory net too
broad relative to the potential risks, and thereby subjects a broad swathe

of developers to unnecessary and onerous informational requirements, which potentially discourages innovation (Rowland 2009, 424). Others criticize the vagueness of the approach, noting that due to the variation in case-by-case assessment criteria, "no seed developer submitting an application package for regulatory approval of a new PNT knows what or how much scientific data are required" (Smyth and McHughen 2008, 219). The net result, some claim, is that confusion with this regulatory concept may impede investment and innovation (National Forum on Seed 2006, 1). Still others maintain that the approach is not rigorous enough, particularly with respect to products derived through biotechnology, and that it allows potentially unsafe species to be freely released into the Canadian environment (Barrett and Abergel 2002). The net effect is that there is substantial uncertainty facing both public and private agricultural genomics developers in Canada, a reality that may dissuade funders from entering this arena, particularly with respect to research conducted at educational institutions or for socially beneficial purposes (see Chapter 4).

In addition to domestic regulations, plants and seeds exported from Canada may be subject to biosafety or trade regulations in nations where the products are marketed or sold. These agreements collectively represent an additional piece of the regulatory portion of the IP–Regulatory Complex applicable to novel products of agricultural genomics. Further perspectives on the Canadian PNT regulatory regime are provided in Chapters 3 and 8.

IP Protection for Novel Plant Varieties in Canada

The practical impact of the international IP regime on developers is defined by national laws in the countries where such products may be developed or sold. In Canada, both plant variety protection and patent rights may be available to developers of novel plant varieties where requisite conditions are met. As in the international arena, each of these approaches provides a different avenue and scope of IP protection that must be carefully considered and balanced as elements of a complex against the prerequisites of the different forms of IP protection, the relevant regulatory regimes, and the relevant IP protections that may be available in other countries.

Plant Breeders' Rights Act

In Canada, the *Plant Breeders' Rights Act* (PBRA) provides protection for new varieties of plants. The PBRA was based on the model established by the 1978 version of the UPOV Convention, discussed earlier (Forge 2005), to which Canada became a party in 1991.

Consistent with the UPOV Convention, the PBRA provides a plant breeder with the exclusive right to (among other things) produce and sell reproductive material of a new variety for a period of up to twenty or twenty-five years, depending on the category of the plant (ss. 5 and 6; Forge 2005).[15] Not all varieties qualify for this protection, however, as the PBRA provides that plant variety protection extends only to those varieties belonging to prescribed categories, and found to be *new* varieties, among other things (s. 4[1], emphasis added). The prescribed categories set out in Schedule 1 of the *Plant Breeders' Rights Regulations* identify the types of plants to which the PBRA and associated regulations apply, and run the gamut from agricultural crops (soybean, potato, wheat, and so on) to ornamental crops (rose, chrysanthemum, poinsettia, and so on). Notably, there is also a catch-all category, holding that "all other categories of the plant kingdom, except algae, bacteria, and fungi" are included within the schedule, and hence subject to the PBRA and its regulations (Schedule 1, Category 40). A "new variety" is one that, by reason of one or more identifiable characteristics, is clearly distinguishable from all varieties commonly known at the date of application, is stable in its essential characteristics over each reproductive cycle, and is a sufficiently homogeneous variety.[16] In order to qualify for protection, a new variety may not have been sold inside or outside of Canada any earlier than a prescribed period prior to the filing of an application (s. 4[3]); this has been termed the "no prior commercialization requirement" (Derzko 1994, IV.A).

In Canada, there are a number of limitations on the scope of plant variety protections granted. These include: (1) a research (breeders') exemption whereby protected varieties may be used for breeding and developing new plant varieties; and (2) a farmers' privilege whereby farmers may save and plant seed on their own land (CFIA 2010, s. 9). These exemptions are framed quite broadly, in line with the corresponding UPOV Convention exemptions, and limit the reach-through rights available to developers (and the relative economic value of such rights) significantly. The importance of these exemptions is particularly apparent when these rights are compared with patent rights.

Patents
In recent years, plant breeders' rights have come to coexist in certain jurisdictions with patent rights over plants or portions thereof, creating a complex and sometimes uncertain IP landscape. In some jurisdictions, notably the United States (where patents are called "utility patents" to distinguish them

from so-called plant patents available in that jurisdiction), Japan, and Australia, a developer may seek both patent protection and plant variety protection. In Canada, the situation is complicated by uncertain case authority on the subject of patented plants and portions thereof.

Under Canadian law, a patent is a right to exclude others from making, selling, or using an "invention" (*Patent Act*, s. 42). The classic rationale for granting of a patent is that it attempts to strike a "bargain," whereby an inventor is given a limited-term monopoly in exchange for public disclosure of his or her invention (*Apotex Inc. v. Wellcome Foundation Ltd.*, para. 3). Patents are granted for a period of twenty years, and, to be patentable, the invention must be novel and non-obvious and must have utility. Unlike in the United States, where the *Plant Patent Act*, 35 U.S.C. §15, extends patent protection to asexually reproduced varieties of plants, no *sui generis* plant patent legislation exists in Canada, and inventors looking to secure patent rights over plants are required to turn to the *Patent Act*.

The *Patent Act*, however, is silent as to whether a plant can be patented, and Canadian courts have been repeatedly called on to determine whether a plant or seed amounts to an "invention" within the scope of section 2 of the act. In the 1989 case of *Pioneer Hi-Bred Ltd. v. Canada (Commissioner of Patents)*, the Federal Court of Appeal held that a new soybean variety, developed by artificial crossbreeding but cultivated naturally, was not an invention.[17] In so doing, the court rejected the approach adopted by the US Supreme Court in cases such as *Ex parte Hibberd*, where the US courts acknowledged that hybrid seeds and plants were patentable. Similarly, the majority of the Supreme Court of Canada, in the 2002 case of *Harvard College v. Canada (Commissioner of Patents)*, which involved a patent application for a genetically modified mouse, suggested in *obiter dicta* that the silence of the *Patent Act* with respect to plants, coupled with the existence of the PBRA and the exception from patentability of plants and animals (as higher life forms) in international agreements, was evidence of Parliament's intent that plants not be considered inventions.

However, in the subsequent decision of *Monsanto Canada Inc. v. Schmeiser*, the Supreme Court of Canada opened the door to the de facto patenting of plants in Canada. The majority held that Monsanto's patent did not offend the prohibition against the patenting of higher life forms as the patent in issue related only to a *component* of the plant (a herbicide resistance gene) and not the plant itself, but did not hesitate to extend the patent to every copy of the herbicide resistance gene in a living plant. Indeed, in her dissent, Justice Arbour noted that the majority's reasoning allowed Monsanto to "do indirectly what Canadian patent law has not allowed them to do directly: namely, to acquire patent protection over whole plants" (para. 108, citing Gold and Adams 2001).

The state of patent law applicable to plants is thus somewhat uncertain in Canada. Whereas higher life forms are not patentable, the same is not true of the individual components of these organisms, such as a gene or cell. The fact that a gene may be expressed in every cell of a higher life form does not seem to factor into this analysis, which apparently allows developers to claim patent protection on all cells or genes within a plant and gain de facto protection over the entire plant.

This uncertainty adds an extra layer to the IP–Regulatory Complex as it pertains to IP in Canada. Indeed, depending on the variety and the novelty of the advancement, owners may opt to seek plant variety protection or, if framed according to the *Monsanto* decision, patent protection (Durell 2006), or both plant variety protection over the variety and patent protection over a modified gene or cell. The impact of this choice goes beyond the mere differences in duration of protection provided. Indeed, as detailed above, the form and substance of the exclusive rights created by the two regimes, as well as the means of their acquisition, differ significantly. More narrowly, the two IP regimes also differ significantly in their respective treatment of the concept of farmers' rights, which allow farmers to save seeds from protected varieties without infringing property rights. Farmers' rights are contemplated under the UPOV Convention but are ultimately left to national policy makers to elect. In *Monsanto*, the Supreme Court of Canada addressed the inconsistency between farmers' rights to save and reuse seed, which *is* included as a part of plant variety protection in Canada, and the *Patent Act*, which allows for no such right (Rimmer 2008, 71). Despite calls by some law reform groups for the introduction of such an exemption from infringement into the *Patent Act* (Canadian Biotechnology Advisory Committee 2002), the Supreme Court held that the *Patent Act* contained no provision allowing farmers to save and reuse seed, and that such a question belonged more in the realm of property rights than patent rights (*Monsanto*, para. 96).

Canadian plant variety protection and patent protection regimes also differ in the scope of available research exemptions. Plant variety protection in Canada affords breeders the right to use protected varieties for breeding and developing new plant varieties without the permission of the rights holder.

In contrast, the scope and robustness of the research exemption available under Canadian patent law remains much less clear (for discussions of the research exemption that exists in Canadian patent law, see Goulding et al. 2011; Ferance 2003). The net result depends on the circumstances and the nature of the innovation at issue, but in some circumstances developers may enjoy more flexibility if they choose to pursue plant variety protection but a potentially stronger – if more uncertain – protection of property rights under patent law.

Conclusion

This chapter has reviewed a wide range of biosafety and intellectual property regimes, both in Canada and internationally, that may be applicable to a product of publicly funded agricultural genomics research. The variety of international and national IP regimes that may provide innovators with choices for protecting new plant varieties, and the variety of international and national regulatory regimes that may affect the development, release, and importation of new plant varieties, raise a number of important questions, many of which should be confronted at an early stage of product development.

While the aims and requirements of biosafety and IP regimes may appear disparate, the reality is that the net impact of applicable regimes on product development should in a number of circumstances be considered *in toto*. In this volume, we refer to the IP and regulatory regimes that are or may be relevant to a given agricultural development project as the Intellectual Property–Regulatory Complex applicable to that project, a concept explored in greater detail in later chapters. As an example of difficulties that aspects of the IP–Regulatory Complex may create, a developer of a product stemming from agricultural genomics research would need to consider not only the relevant biosafety regimes that may apply domestically and internationally depending on intended uses, but also the relevance of available property protection or avenues for widespread sharing that may bear on return on investment or mechanisms for uptake. The scope of applicable biosafety regulations may impact considerations of how to share end products or research material and the pursuit of property interests. Similarly, a developer who intends to share a socially beneficial product of agricultural genomics under the ITPGRFA and related regimes (rather than pursue property interests) must consider whether international biosafety hurdles and the costs associated therewith might be prohibitive.

In our own work with a publicly funded genomics research project, we have found that the Canadian PNT regulatory structure has direct implications for a dual-use food/fuel sunflower product. Indeed, it is likely that increased "woodiness" may change the sunflower's competitiveness relative to its wild relatives, and therefore could be deemed to constitute a novel trait that requires an environmental safety assessment prior to its release in Canada.[18] Genomics researchers wishing to develop the woody sunflower in Canada, or export it from Canada, would likely need to make a thorough assessment based on applicable legislation and CFIA policies. Internationally, the requirements of the CBD and the WTO would impose similar requirements in countries interested in importing such a dual-use sunflower. Regulatory regimes in each of those countries would need to be carefully surveyed at an early stage in order to estimate whether the product could be successfully imported in the country or countries of interest within the developers' desired timeline and budget.[19]

In the IP context, the developers of the woody sunflower would have a variety of choices, beginning with plant variety protection in Canada, assuming that the sunflower met the definition of a new variety under the *Plant Breeders' Rights Act*. Furthermore, if developed within the landscape of patentable subject matter as modified by the *Pioneer Hi-Bred, Harvard*, and *Monsanto* decisions, developers may have a form of patent protection available to them.[20] The ability to claim plant variety protection, patent protection, or other exclusive rights over the sunflower in other jurisdictions of interest internationally would be highly dependent on the jurisdictions at issue and the IP management strategies chosen by the developers. For example, both patent and plant variety protection might be sought in countries such as the United States, Japan, and Australia, but in other countries, including developing nations where the sunflower might be most helpful but farmers' rights and other proprietary rights exceptions are likely to be strongest, choices would have to be made. In circumstances where the ultimate goal of a project is the dissemination of seeds or other agricultural products to those in need, rather than the commercial enjoyment of the resulting products, developers may wish to take advantage of seed-sharing mechanisms such as the ITPGRFA rather than the proprietary protections available in jurisdictions of interest, but such matters are highly circumstance-dependent. IP management strategies, and the importance of tailoring them to particular jurisdictions and their applicable sociological considerations, are the focus of Chapter 7.

In conclusion, we have found that there is little appreciation for the multifaceted aspects of the IP–Regulatory Complex, with stakeholders often considering biosafety independently from intellectual property and sharing (if biosafety or IP are considered at all). We believe that a lack of attention to the IP–Regulatory Complex could limit the appreciation of both practical barriers to innovation and delivery of new products and ultimately the impact of the products. In light of this reality, we urge further discussion on mechanisms to apprehend and address the IP–Regulatory Complex in a manner that facilitates beneficial applications of genomics research. As governments and publicly funded agencies place greater focus on returns from investment in research (financial and otherwise), new mechanisms will be more important for both genomics researchers and the ultimate beneficiaries of product-driven research.

Notes

1 In part, this may be because the products that emerge from such funding efforts defy convenient regulatory characterization in existing categories, as either genetically modified organisms or conventionally bred products. There are no uniform conceptions of "genetically modified" organisms or "living modified" organisms among international regulators, and the term encompasses a wide variety of modification techniques and approaches to product development. In this chapter, the use of terms such as "GM" and "GMOs" will thus be carefully defined and used in the context of particular national and international definitions, and will not be limited to a singular conceptual definition.

2 This information is obtained using the calculator available at FAOStat (http://www.faostat.fao.org), and calculating the "Gross Production Value" throughout the "World" for "Sunflower Seed."

3 The *Agreement on Technical Barriers to Trade,* April 15, 1994, 1186 U.N.T.S. 276 (entered into force January 1, 1995), serves a similar purpose, but the SPS Agreement takes precedence where the measure taken meets the definition of a "sanitary or phytosanitary procedure" defined in Annex A(1) as, among other things, "any measure applied ... to protect animal or plant life or health within the territory of the Member ... [or] to prevent or limit other damage within the territory of the Member from the entry, establishment or spread of pests." Genetically modified organisms' (GMO) movement and biosafety are clearly within the scope of the SPS Agreement (see *Panel Report, European Communities – Measures Affecting the Approval and Marketing of Biotech Products*). Biosafety measures with respect to non-GMO plant materials are more clearly within the scope of the SPS Agreement than the *Agreement on Technical Barriers to Trade,* and the application of the latter will be largely ignored in this chapter. Measures covered by the SPS Agreement include, pursuant to Annex A, "all relevant laws, decrees, regulations, requirements and procedures including, *inter alia,* end product criteria; processes and production

methods; testing, inspection, certification and approval procedures; quarantine treatments ... and packaging and labeling requirements directly related to food safety."

4 Defined as "any living organism that possesses a novel combination of genetic material obtained through the use of modern biotechnology" (Article 3[g]; see Secretariat of the CBD 2000). "Living organism," in this context, refers to "any biological entity capable of transferring or replicating genetic material, including sterile organisms, viruses and viroids," and "biotechnology" is defined as the application of techniques that "overcome natural physiological reproductive or recombination barriers and that are not techniques used in traditional breeding and selection" (Article 3[h], 3[i]; see Secretariat of the CBD 2000).

5 The *Cartagena Protocol* directs that LMOs should be assessed on a case-by-case basis, risks should be evaluated in the context of the likely receiving environment, and "lack of scientific knowledge or scientific consensus should not necessarily be interpreted as indicating a particular level of risk, an absence of risk, or an acceptable risk" (Annex III: Risk Assessment, s. 4; Secretariat of the CBD 2000).

6 Prior to such legislative enactments, plant breeders to some extent engineered an earlier form of exclusivity with the development of hybrid plants, which cannot be replanted or propagated from saved seed.

7 The different revisions of the UPOV Convention vary on this latter exception: the 1978 version is silent on "farmers' privilege" but refers to the right of farmers to use seed harvested from protected varieties for private and noncommercial purposes, effectively acknowledging the practice, whereas the 1991 version explicitly leaves farmers' privilege to national legislation, but notes that member states must take measures to safeguard the "legitimate interests of the breeder" even in instituting farmers' rights protections.

8 See, e.g., *Diamond v. Chakrabarty,* in which the US Supreme Court proclaimed that patentable subject matter was to include "anything under the sun that is made by man." In certain jurisdictions, precedents such as *Chakrabarty* have led to findings that new plant varieties are patentable subject matter.

9 The potential for inconsistent national regimes has been recognized in other international conventions. For example, the *Convention on Biological Diversity* explicitly acknowledges this potential issue, noting in Article 16(5) that "the Contracting Parties, recognizing that patents and other intellectual property rights may have an influence on the implementation of this Convention, shall cooperate in this regard subject to national legislation and international law in order to ensure that such rights are supportive of and do not run counter to its objectives." This provision has never been called on to resolve such issues.

10 See, e.g., the Access and Benefit Sharing (ABS) provisions of the *Convention on Biological Diversity,* and the relevant provisions of the *Nagoya Protocol* of October 29, 2010.

11 "Intended for consumption as food": Novel products intended for consumption as food (i.e., those that do not have a safe history of use as a food, were developed by a process not previously applied to that food, have undergone a major change, or have been genetically modified) are regulated under the *Food and Drug Regulations,* C.R.C., c. 870, Part B. Foods are compared to appropriate analogues having a history

of safe use, with any significant differences in allergenicity, toxicity, or nutritional quality being assessed for potential adverse health effects. It is our view that these coordinate regimes may add further procedural hurdles to commercialization and/ or development of novel agricultural products in the case of repeated requests for the provision and scientific analysis of additional data (Smyth and McHughen 2008).

"All novel plants developed within Canada": Such an approach is consistent with Canada's obligations under the *Convention on Biological Diversity* to regulate the use and release of any organisms that are likely to have adverse environmental impacts (Article 8) and to assess the environmental impact of potentially hazardous products prior to their release.

12 This novelty assessment is characterized as a form of "pre-regulation," where the novelty of a plant must be disproven to avoid further regulation by the Plant Biosafety Office (PBO) within the Canadian Food Inspection Agency (CFIA) (Abergel and Barrett 2002, 147–48). The information required by the PBO to assess the safety of a novel plant is significant. It includes the identity and origin of the Plant with Novel Traits, properties of its novel gene products, if any, and a summary of anticipated relative impacts resulting from its release.

13 "Familiarity" is defined as "knowledge of the characteristics of a plant species and experience with the use of that plant species in Canada" (CFIA 2012c). "Substantial equivalence" is defined as "the equivalence of a novel trait within a particular plant species, in terms of its specific use and safety to the environment and human health, to those in that same species, that are in use and generally considered as safe in Canada, based on valid scientific rationale" (ibid.).

14 In making this comparison, the grower should make reference to species-specific companion biology documents, which provide baseline data on approved analogues for many species (CFIA 2013). These data may be obtained through the use of "confined field trials" designed to minimize unnecessary environmental exposures (CFIA 2000, s. 2.1). Confined trials are conducted by developers, and conditions placed on them can vary significantly depending on the plant and novel traits in question, but often include reproductive and physical isolation from other plants, use of a limited area, and post-trial monitoring: See *ibid.*, Appendix III.

15 It is important to note that plant breeders' rights extend only to the exclusive right to sell *reproductive* (propagating) material. See Derzko 1994, noting that s. 5 of the PBRA grants the exclusive right to sell the seed, cuttings, or parts of the plant if the plant reproduces vegetatively. The rights do not extend to the plant itself. This allows farmers to buy the seeds from the breeder, grow the plant, and sell it, without being liable for infringement of the plant breeder's right.

16 "Known at the date of application": Section 5 of the *Plant Breeders' Rights Regulations* states that when determining whether a plant variety is a matter of "common knowledge," according to s. 4(2)(a) of the PBRA, the following criteria should be considered: (a) whether the variety is already being cultivated or exploited for commercial purposes; or (b) whether the variety is described in a publication that is accessible to the public.

A "sufficiently homogeneous variety" is defined in subsection 4(3) of the PBRA, meaning "such a variety that, in the event of its sexual reproduction or vegetative propagation in substantial quantity, any variations in characteristics of plants so reproduced or propagated are predictable, capable of being described and commercially acceptable." See also Ludlow 1993, describing the enactment of the PBRA and the extent of the rights conferred by it; and CFIA 2010.

17 Note that in the appeal heard by the Supreme Court of Canada, Justice Lamer (as he then was) dismissed *Pioneer's* appeal on the basis that the patent claims had failed to meet the disclosure requirements of the *Patent Act,* and abstained from answering the question of whether plant or seed was patentable.

18 "Competitiveness relative to its wild relatives": Wild sunflowers are represented by the PBO in CFIA 2012b. Any concerns of increased or decreased competitiveness would likely be amplified by the fact that the sunflower is a crop for which Canada is a "centre of diversity," and therefore warrants additional scrutiny: see CFIA 2009, s. 2.3.2(3).

"Constitute a novel trait": Of the three decision documents published by the CFIA that deal with sunflowers (DD2010-80, DD2008-69, and DD2005-50), none has explicitly mentioned woodier sunflowers, and presumably the matter has not yet been decided upon by the PBO. The decision documents are available publicly: CFIA 2013.

"Environmental safety assessment": It is worth noting that *in practice,* seeds can be made available outside Canada without triggering the PNT regulations by placing the germplasm in a gene bank, such as the National Plant Germplasm System (NPGS) of the US Department of Agriculture and the Consultative Group for International Agricultural Research (CGIAR) research centres. See Bretting 2007, 760. To a large degree, these centres have become increasingly important as a balance to the simultaneous proliferation of property rights in plant resources, as discussed in greater detail below.

19 Depending on the ratification status of the *Cartagena Protocol* in the importing nation(s), the means by which the sunflower was developed could also significantly affect the time and costs of regulatory approval (in the sense that a new sunflower product classified as an LMO could, as such, attract additional regulatory scrutiny).

20 The above discussion on patentable subject matter relates to claims directed to the plant itself, of course, and should not be read as derogating from the right of the inventors to claim patent protection over, for example, new and non-obvious processes or business methods relating to the development of the sunflower.

References

Legislation, Treaties, and Agreements

Agreement on the Application of Sanitary and Phytosanitary Measures, April 15, 1994, 1867 U.N.T.S. 493 (entered into force January 1, 1995) [SPS Agreement].

Agreement on Trade-Related Aspects of Intellectual Property Rights, April 15, 1994, *Marrakesh Agreement Establishing the World Trade Organization,* Annex 1C, Legal Texts: The Results of the Uruguay Round of Multilateral Trade Negotiations 320 (1999), 1869 U.N.T.S. 299, 33 I.L.M. 1197 [TRIPS Agreement].

Convention on Biological Diversity, June 5, 1992, 1760 U.N.T.S. 79, U.N. Doc. ST/DPI/1307, 31 I.L.M. 818 (entered into force December 29, 1993) [CBD].

Food and Drug Regulations, C.R.C., c. 870.

Food and Drugs Act, R.S.C. 1985, c. F-27.

General Agreement on Tariffs and Trade, October 30, 1947, 55 U.N.T.S. 187, Can. T.S. 1947 No. 27 (entered into force January 1, 1948) [GATT].

International Convention for the Protection of New Varieties of Plants, December 2, 1961, 815 U.N.T.S. 89 (as revised at Geneva on November 10, 1972, on October 23, 1978, and on March 19, 1991) [UPOV Convention].

International Treaty on Plant Genetic Resources for Food and Agriculture, adopted November 3, 2001. S. Treaty Doc. No. 110-19 [ITPGRFA]. ftp://ftp.fao.org/docrep/fao/007/ae040e/ae040e00.pdf.

International Treaty on Plant Genetic Resources for Food and Agriculture, Standard Material Transfer Agreement, adopted June 16, 2006, Resolution 1/2006 of the Governing Body [SMTA].

Nagoya Protocol on Access to Genetic Resources and the Fair and Equitable Sharing of Benefits Arising from Their Utilization to the Convention on Biological Diversity, Conference of the Parties to the Convention on Biological Diversity, UNEP/CBD/COP/DEC/X/1 of October 29, 2010. [Nagoya Protocol].

Panel Report, European Communities – Measures Affecting the Approval and Marketing of Biotech Products, WTO Doc WT/DS291/R, WT/DS292/R, WT/DS293/R (September 29, 2006).

Patent Act, R.S.C., 1985, c. P-4.

Plant Breeders' Rights Act, S.C. 1990, c. 20 [PBRA].

Plant Breeders' Rights Regulations, SOR/91 594.

Plant Patent Act, 35 U.S.C. § 15.

Seeds Act, R.S.C., 1985, c. S-8.

Seeds Regulations, C.R.C., c. 1400, Part 5.

Cases

Apotex Inc. v. Wellcome Foundation Ltd., 2002 SCC 77, [2002] 4 S.C.R. 153.

Appellate Body Report, EC Measures Concerning Meat and Meat Products (Hormones) (January 16, 1998), WTO Doc WT/DS26/AB/R.

Diamond v. Chakrabarty, 447 U.S. 303 (1980).

Ex parte Hibberd, [227] U.S.P.Q. 443 (Bd. Pat. App. 1985).

Harvard College v. Canada (Commissioner of Patents), 2002 SCC 76, [2002] 4 S.C.R. 45 [*Harvard*].

Monsanto Canada Inc. v. Schmeiser, 2004 SCC 34, [2004] 1 S.C.R. 902 [*Monsanto*].

Pioneer Hi-Bred Ltd. v. Canada (Commissioner of Patents), [1987] F.C.J. No. 238, [1987] 3 F.C. 8 (C.A.).

Secondary Sources

Abergel, Elisabeth. 2007. "Trade, Science, and Canada's Regulatory Framework for Determining the Safety of GE Crops." In *Genetically Engineered Crops: Interim Policies, Uncertain Legislation,* ed. Iain E.P. Taylor, 173–206. New York: Haworth Food and Agricultural Products Press.

Abergel, Elisabeth, and Katherine Barrett. 2002. "Putting the Cart before the Horse: A Review of Biotechnology Policy in Canada." *Journal of Canadian Studies/Revue d'etudes canadiennes* 37 (3): 135–61.

Arias, D.M., and Loren H. Rieseberg. 1994. "Gene Flow between Cultivated and Wild Sunflowers." *Theoretical and Applied Genetics* 89 (6): 655–60. http://dx. doi.org/10.1007/BF00223700.

Barrett, Katherine, and Elisabeth Abergel. 2000. "Breeding Familiarity: Environmental Risk Assessment for Genetically Engineered Crops in Canada." *Science and Public Policy* 27 (1): 2–12. http://dx.doi.org/10.3152/147154300781782138.

–. 2002. "Defining a Safe Genetically Modified Organism: Boundaries of Scientific Risk Assessment." *Science and Public Policy* 29 (1): 47–58. http://dx.doi.org/10.3152/147154302781781128.

Bretting, Peter. 2007. "The U.S. National Plant Germplasm System in an Era of Shifting International Norms for Germplasm Exchange." *Acta Horticulturae* 760 (5): 55–60.

Canadian Biotechnology Advisory Committee. 2002. Patenting of Higher Life Forms and Related Issues. Government of Canada Publications, http://publications. gc.ca/collections/Collection/C2-598-2001-2E.pdf.

Canadian Food Inspection Agency. 2000. *Directive 2000–07: Conducting Confined Research Field Trials of Plants with Novel Traits.* Ottawa: CFIA.

–. 2004. *Directive 94–08: Assessment Criteria for Determining Environmental Safety of Plants with Novel Traits.* Ottawa: CFIA.

–. 2009. *Directive 2009–09: Plants with Novel Traits Regulated under Part V of the Seeds Regulations: Guidelines for Determining When to Notify the CFIA.* Ottawa: CFIA.

–. 2010. *Guide to Plant Breeders' Rights.* Canadian Food Inspection Agency, http://www.inspection.gc.ca/english/plaveg/pbrpov/guidee.shtml.

–. 2012a. *2011–2012 Departmental Performance Report.* Canadian Food Inspection Agency, http://www.inspection.gc.ca/about-the-cfia/accountability/reports-to -parliament/2011-2012-dpr/eng/1348777953917/1348778053447?chap=3#s1c3.

–. 2012b. *Biology Document BIO2005–01: The Biology of* Helianthus annuus L. Canadian Food Inspection Agency, http://www.inspection.gc.ca/plants/plants -with-novel-traits/applicants/directive-94-08/biology-documents/helianthus -annuus-l-/eng/1330977236841/1330977318934.

–. 2012c. *The Biology of* Zea mays *(L.).* Canadian Food Inspection Agency, http:// www.inspection.gc.ca/plants/plants-with-novel-traits/applicants/directive-94 -08/biology-documents/zea-mays-l-/eng/1330985739405/1330985818367#A1.

–. 2013. "Decision Documents – Determination of Environmental and Livestock Feed Safety." Canadian Food Inspection Agency, http://www.inspection.gc.ca/ plants/plants-with-novel-traits/approved-under-review/decision-documents/ eng/1303704378026/1303704484236.

Carrizosa, Santiago, Stephen B. Brush, Brian D. Wright, et al., eds. 2004. *Accessing Biodiversity and Sharing the Benefits: Lessons from Implementing the Convention on Biological Diversity.* UCN Environmental Policy and Law Paper No. 54. Gland, Switzerland: International Union for Conservation of Nature and Natural Resources.

Derzko, Natalie. 1994. "Plant Breeders' Rights in Canada and Abroad: What Are These Rights and How Much Must Society Pay for Them?" *McGill Law Journal/Revue de droit de McGill* 39 (1): 144–78.

Durell, Karen. 2006. "Functional Foods and Intellectual Property Rights: The Importance of an Integrated Approach." *Health Law Review* 15 (1): 14–22.

Dutfield, Graham. 2008. "Turning Plant Varieties into Intellectual Property: The UPOV Convention." In *The Future Control of Food: A Guide to International Negotiations and Rules on Intellectual Property, Biodiversity and Food Security,* ed. Geoff Tansey and Tasmin Rajotte, 27–47. London: Earthscan.

Ferance, S.J. 2003. "The Experimental Use Defence to Patent Infringement." *Canadian Intellectual Property Review* 20 (1): 1.

Food and Agriculture Organization of the United Nations (FAO). 2014. FAOStat, http://faostat3.fao.org/home/E.

Forge, Frédéric. 2005. "Intellectual Property Rights in Plants and the Farmer's Privilege." Library of Parliament, Parliamentary Information and Research Service.

Genome British Columbia. 2013. "Genomics of Sunflower." Genome BC, http://www.genomebc.ca/portfolio/projects/bioenergy-projects/genomics-of-sunflower/.

Gold, E.R., and W.A. Adams. 2001. "The *Monsanto* Decision: The Edge or the Wedge." *Nature Biotechnology* 19 (6): 587. http://dx.doi.org/10.1038/89355.

Goulding, Rebecca, Emily Marden, Rachael Manion, et al. 2010. "Alternative Intellectual Property for Genomics and the Activity of Technology Transfer Offices: Emerging Directions in Research." *Boston University Journal of Science and Technology Law* 16 (2): 194–230.

Goulding, Rebecca, Matthew Voell, Emily Marden, et al. 2011. "Expansion of the Canadian Research Exemption for Biotechnology Research Tools." *Biotechnology Law Report* 30 (1): 59–63. http://dx.doi.org/10.1089/blr.2011.9983.

Halewood, Michael, and Kent Nnadozie. 2008. "Giving Priority to the Commons: The International Treaty on Plant Genetic Resources for Food and Agriculture (ITPGRFA)." In *The Future Control of Food: A Guide to International Negotiations and Rules on Intellectual Property, Biodiversity and Food Security,* ed. Geoff Tansey and Tasmin Rajotte, 115–40. London: Earthscan.

Hope, Janet. 2008. *Biobazaar: The Open Source Revolution and Biotechnology.* Cambridge, MA: Harvard University Press.

Llewelyn, Margaret. 1997. "The Legal Protection of Biotechnological Inventions: An Alternative Approach." *European Intellectual Property Review* 19 (3): 115–27.

Ludlow, Gregory. 1993. "Intellectual Property Law (1987–93) Part I – Summary of Government Activity." *Ottawa Law Review* 25 (1): 89–122.

Marden, Emily. 2003. "Risk and Regulation: U.S. Regulatory Policy on Genetically Modified Food and Agriculture." *Boston College Law Review* 44 (3): 733–88.

Marden, Emily, and Godfrey R. Nelson. 2012. "Intellectual Property and Sharing Regimes in Agricultural Genomics: Finding the Right Balance for Innovation." *Drake Journal of Agricultural Law* 17 (2): 369–93.

National Forum on Seed. 2006. *Plants with Novel Traits Working Group 1 Summary Report.* Ottawa: National Forum on Seed.

National Science Foundation. 2010. "Press Release 10-201: NSF Awards New Projects for Plant Genome Research." National Science Foundation, http://www.nsf.gov/news/news_summ.jsp?cntn_id=117768.

Oguamanam, Chidi. 2006. "Intellectual Property Rights in Plant Genetic Resources: Farmers' Rights and Food Security of Indigenous and Local Communities." *Drake Journal of Agricultural Law* 11 (3): 273–306.

Phillips, Peter, and William Kerr. 2000. "The WTO versus the Biosafety Protocol for Trade in Genetically Modified Organisms." *Journal of World Trade* 34 (4): 63–75.

Pray, C.E., and A. Naseem. 2003. "The Economics of Agricultural Biotechnology Research." Food and Agriculture Organization of the United Nations ESA Working Paper No. 03-07.

Rajotte, Tasmin. 2008. "The Negotiations Web: Complex Connections." In *The Future Control of Food: A Guide to International Negotiations and Rules on Intellectual Property, Biodiversity and Food Security,* ed. Geoff Tansey and Tasmin Rajotte, 141–68. London: Earthscan.

Rimmer, Matthew. 2008. *Intellectual Property and Biotechnology: Biological Inventions.* Cheltenham, UK: Edward Elgar. http://dx.doi.org/10.4337/9781848440180.

Roffe, Pedro. 2008. "Bringing Minimum Global Intellectual Property Standards into Agriculture: The Agreement on Trade-Related Aspects of Intellectual Property Rights (TRIPS)." In *The Future Control of Food: A Guide to International Negotiations and Rules on Intellectual Property, Biodiversity and Food Security,* ed. Geoff Tansey and Tasmin Rajotte, 48–68. London: Earthscan.

Rowland, G.G. 2009. "The Effect of Plants with Novel Traits (PNT) Regulation on Mutation Breeding in Canada." In *Induced Plant Mutations in the Genomics Era,* ed. Q.Y. Shu, 423–24. Rome: Food and Agriculture Organization of the United Nations.

Safrin, Sabrina. 2002. "Treaties in Collision? The Biosafety Protocol and the World Trade Organization Agreements." *American Journal of International Law* 96 (3): 606–28. http://dx.doi.org/10.2307/3062164.

Secretariat of the CBD. 2000. *The Cartagena Protocol on Biosafety to the Convention on Biological Diversity: Text and Annexes.* Montreal: CBD.

Smyth, Stuart, and Alan McHughen. 2008. "Regulating Innovative Crop Technologies in Canada: The Case of Regulating Genetically Modified Crops." *Plant Biotechnology Journal* 6 (3): 213–25. http://dx.doi.org/10.1111/j.1467-7652.2007.00309.x.

Snow, Allison, Pedro Moran-Palma, Loren H. Rieseberg, et al. 1998. "Fecundity, Phenology, and Seed Dormancy of F_1 Wild-Crop Hybrids in Sunflower (*Helianthus annuus,* Asteraceae)." *American Journal of Botany* 85 (6): 794–801. http://dx.doi.org/10.2307/2446414.

Thorpe, Phil. 2002. *Study on the Implementation of the TRIPS Agreement by Developing Countries.* London: Commission on Intellectual Property Rights.

Van Overwalle, Geertrui, ed. 2009. *Gene Patents and Collaborative Licensing Models: Patent Pools, Clearinghouses, Open Source Models and Liability Regimes.* Cambridge: Cambridge University Press. http://dx.doi.org/10.1017/CBO9780511581182.

Waltz, Emily. 2010. "Plant Genomics' Ascent." *Nature Biotechnology* 28 (1): 10. http://dx.doi.org/10.1038/nbt0110-10b.

2

The Treatment of Social and Ethical Concerns in Regulatory Responses to Agricultural Biotechnology

A Historical Analysis

SARAH HARTLEY

Science and technology represent one of the pillars of economic growth in the industrialized world. Yet, while technological developments fuel the domestic economy and international competitiveness, governments increasingly face public policy decisions about the risks and desirability of new technologies as commercial products emerge on the market. Thus, advances in agricultural biotechnology in the 1980s triggered the development of policy responses to human health and environmental risks and led to important large-scale public and political debates about the social and ethical implications of biotechnology that continue to impact current developments in genomics.

Since the 1990s, a body of literature on the social and ethical issues related to agricultural biotechnology (including genetics and genomics) has developed, exploring the significance of such social and ethical issues for governance (Nuffield Council on Bioethics 2012; Brunk and Hartley 2012; Federal Ethics Committee on Non-Human Biotechnology 2012; Bunton and Peterson 2005). These issues include the distribution of risks and benefits, the impact on agricultural production systems and communities, transparency, corporate control of the food production systems, global inequalities between the "overdeveloped" north and the underdeveloped south, the challenges of food insecurity and the risks of biodiversity loss, the role and direction of technological progress, and, ultimately, the desirability of the technology (Nuffield Council on Bioethics 2012; Phillips et al. 2010; Dowdeswell, Daar, and Singer 2005).

The term "agricultural biotechnology" is often used in the literature and the media to describe genetic modification; however, it is a broader term that for the purpose of this chapter is understood to include genomics, proteomics, and cloning. Some scholars have described developments in agricultural biotechnology as a transition from genetic technologies, particularly genetically modified (GM) crops, to next-generation "omics" technologies, such as genomics and proteomics (Laycock and Howlett 2012; Gottweis 2005). This transition triggered investigations into the implications of emerging next-generation technologies for governance frameworks (Howlett and Laycock 2012). David Laycock and Michael Howlett (2012) argue that next-generation policies and regulations need to rely on comparative historical analyses of the impacts and results of early policy and regulatory frameworks. This chapter contributes to this endeavour by providing a historical analysis of the treatment of the social and ethical concerns relating to agricultural biotechnology in the development of the initial government response thereto, using Canada as a case study. It was technological developments in genetically modified crops and the impending commercialization of these technologies that triggered a policy and regulatory response to agricultural biotechnology in Canada; hence, the debate about agricultural biotechnology was essentially a debate about GM crops.

Importantly, these historical developments continue to form the foundation of the regulatory framework for products of agricultural biotechnology in Canada, and are therefore applicable to the development of products from agricultural genomics: the regulatory framework for agricultural biotechnology applies to all products that have a "novel trait," whether developed using GM technologies or using genomics. Thus, just as social and ethical considerations were excluded from the regulatory framework for the early products of biotechnology, they continue to be outside governance considerations in the regulatory framework for genomics.

Through an in-depth analysis of policy documents, reports, minutes, and interviews with senior public officials, this chapter traces the emergence, shaping, and treatment of social and ethical concerns raised by agricultural biotechnology during the establishment of its policy and regulatory framework in Canada; it does so by identifying the points in the policy cycle when experts and the lay public raised and discussed the relevant social and ethical issues in the debate about the risks and benefits of GM crops. The chapter focuses on when, where, and by whom these issues were raised, through an examination of public and expert consultations, advisory committees, and commissioned reports, as well as the mechanisms the Canadian government employed to consider these issues explicitly.

This analysis demonstrates the marginal role that social and ethical issues played in the development of the policy and regulatory framework for GM crops (and consequently agricultural biotechnology in Canada generally), and the subordination of these issues to science, due in part to the institutionalization or culturally embedded nature of science. Despite critical junctures at which the Canadian government was challenged to consider the social and ethical issues, the scientific determination of risk continued to dominate the policy and regulatory framework. The social and ethical issues associated with agricultural biotechnology, including agricultural genomics, are little understood and continue to be excluded from debates about risk and governance more broadly.

The chapter tracks chronologically the emergence of social and ethical issues associated with agricultural biotechnology during development of the policy and regulatory framework, drawing attention to public and stakeholder engagement exercises as opportunities to flesh out the issues, and it also explores other ways in which these issues were given recognition. In the late 1990s, the government established the Canadian Biotechnology Advisory Committee (CBAC) and instructed the Royal Society of Canada to consider the human health and environmental issues. The results of these government decisions are analyzed, after which the case of GM wheat is introduced to demonstrate the resilience of the policy and regulatory framework against challenges to consider socio-economic issues in regulatory decisions. The final section discusses the concept of scientism as a tool for understanding the exclusion of social and ethical issues in policy and regulatory frameworks for managing the risks of agricultural biotechnology; it also discusses the relevance of both the Canadian regulatory approach to the Intellectual Property–Regulatory Complex and the "responsible research and innovation" approach to governance recently adopted in some European jurisdictions.

Establishing the Policy and Regulatory Framework
for Agricultural Biotechnology

The social and ethical issues associated with agricultural biotechnology were largely missing from discussions that shaped the regulatory framework. By the mid-1980s, developments in agricultural biotechnology had progressed to the point where GM crops needed to be tested in the field and new environmental risks needed to be assessed through a regulatory framework. Once it became clear that a policy and regulatory response was needed to manage these new environmental risks, the debate over what such a response would look like opened up. Numerous workshops and meetings yielded a plethora of reports and committees were established to investigate and debate the type of regulatory framework that would be required. However, the debate focused largely on the benefits of agricultural biotechnology, particularly the economic benefits, and a narrow set of human health and environmental risks, while social and ethical issues were largely absent.

The Organisation for Economic Co-operation and Development (OECD) shared the Canadian government's excitement over the potential benefits of agricultural biotechnology, promoted the goals of competitiveness and harmonization, and emphasized the importance of existing regulations as a means of achieving these goals. It established a narrative that organisms developed by recombinant DNA (rDNA) techniques posed no unique hazards (Bull, Holt, and Lilly 1982) and, along with other international reports, set the foundations for key scientific concepts in the final regulatory framework (Barrett 1999). The arguments advanced by the OECD resonated with the biotechnology industry and with the recommendations of other reports commissioned by government. Canadian industry representatives, molecular biologists, and senior officials at the Canadian Food Inspection Agency (CFIA) argued that negative consequences arising from the use of agricultural biotechnology were extremely unlikely or nonexistent, and no different from those posed by products developed through conventional agriculture (Scowcroft and Holbrook 1988; Henley 1986; Canadian Agricultural Research Council 1988).

In 1993, the Canadian government established the National Biotechnology Advisory Committee (NBAC), to advise the Minister of State for Science and Technology on new developments and policy requirements in biotechnology. The NBAC and the biotechnology industry preferred a regulatory system that was based on science alone and immune to social and ethical issues. The NBAC played a prominent role in the debate over the requirements for a policy and regulatory framework. Supportive of the biotechnology industry's desire to move ahead with commercialization, it stressed the importance of biotechnology for Canada's economic progress and international competitiveness and recommended that Canada establish a regulatory framework that would rely solely on a scientific assessment of new products and an independent scientific advisory body to provide a single drop-off point for industry applications (NBAC 1988).

The Canadian Agricultural Research Council organized a workshop in 1988 to debate regulatory requirements for the environmental release of GM plants. All one hundred participants came from industry (agricultural and biotechnology), user groups, academia (scientists), and government. No public interest groups, nongovernmental organizations (NGOs), humanists, or social scientists attended. The most significant recommendation from the workshop was that the focus of regulations should be products derived from biotechnology rather than the process that produces them. The workshop report also recommended educating the public about the benefits of bio- technology (Canadian Agricultural Research Council 1988). Social and eth- ical issues were largely absent from the discussion, although concern was raised about the possible conflict of interest in Agriculture Canada's dual role as promoter and regulator (Canadian Agricultural Research Council 1988).

The decision to use existing legislative and policy frameworks to regulate agricultural biotechnology was further institutionalized in the late 1980s by two of the more influential reports: *The Regulation of Plant Biotechnology in Canada* (Caldwell and Duke 1988) and *The Regulation of Plant Biotechnology in Canada: Part 2, The Environmental Release of Genetically Altered Plant Material* (Kalous and Duke 1989). These reports set out the first regulatory framework for biotechnology in Canada (Barrett 1999). R.B. Caldwell and L.H. Duke's recommendations (1988) were based on a study of the state of the technology through a survey of plant researchers who saw rDNA tech- niques as holding great potential for agricultural biotechnology. In their re- port published a year later, M.J. Kalous and L.H. Duke (1989) drew heavily on the assessment of risks presented by Caldwell and Duke (1988) in devel- oping recommendations for a regulatory framework.

By the late 1980s, according to Environment Canada official Brewster Kneen (1992), the social and ethical impacts of agricultural biotechnology were being debated in Europe and a number of questions were raised that were not being considered in Canada. These included concerns about the impact of agricultural biotechnology on overproduction in certain key areas such as chemical and/or food production, existing food surplus, and the economies of developing countries. Questions were also being raised about the patenting of organisms and the potential for large pharmaceut- ical, chemical, and biotechnology companies to control the agricultural sec- tor. At this time, the Canadian Institute for Environmental Law and Policy began to raise objections to the exclusion of social and ethical issues from the regulatory debate, and concerns about the narrow conceptualization of science (CIELAP 1988).

The *Canadian Environmental Protection Act* (CEPA) was enacted in 1988. Its purpose was to regulate biotechnology products not regulated under other acts, including aquatic organisms, livestock, or production organisms derived through biotechnology (CFIA 2014). The enactment of CEPA provided an opportunity for public discussion of agricultural bio-technology and the attendant social and ethical issues. Environment Canada organized a public consultation exercise involving over 1,600 Canadians (NBAC 1988). This workshop flushed out concerns from the environmental community with respect to the scientific determination of environmental risks (McIntyre 1990), but social and ethical issues remained absent from the workshop and there was a lack of public interest, with extremely low public turnout at the national consultation meetings sponsored by Environment Canada (McIntyre 1990).

Pressure to establish a regulatory framework for agricultural biotechnol-ogy grew in the early 1990s as GM crops neared commercialization. The government needed to find a way to balance the competitiveness of the in-dustry with protecting the environment and human health from the risks associated with commercial planting. By 1992, the Canadian government had approved over six hundred field trials for GM crops; another five hun-dred trials were approved the following year (CFIA 2003).

In December 1992, Cabinet decided to establish a federal regulatory framework and develop guidelines for evaluating the products of biotech-nology. The Federal Framework for Biotechnology was formally announced through the Interdepartmental Committee on Biotechnology in January 1993 (AAFC, HC, and EC 1993). The framework is science-based and offers no opportunity for social and ethical issues to influence risk decision making. Health and Welfare Canada, Agriculture Canada, and Environ-ment Canada developed the guiding principles, defined as:

- science-based risk assessment
- protection of human health and the environment
- use of existing legislation and regulatory institutions
- harmonization with national priorities and international standards
- assessment of product, not process
- openness and consultation
- investment facilitation, development innovation, and adoption of sus-tainable Canadian biotechnology products and processes.

Genetically modified plants – or any other novel plants developed from agricultural biotechnology – were to be regulated under the *Seeds Act,* microbial feed supplements under the *Feeds Act,* microbial growth supplements under the *Fertilizers Act,* microbial pest control products under the *Pest Control Products Act,* and veterinary vaccines and biologics under the *Health of Animals Act.* The use of existing legislation was officially justified on the basis of efficiency and cost-effectiveness, as well as claims that it built on the experience of previous regulatory activities.

In 1993, Agriculture Canada, Health Canada, and Environment Canada held a series of consultation exercises across the country that culminated in a Workshop on Regulating Agricultural Products of Biotechnology (AAFC, HC, and EC 1993). Although the Federal Framework was already established, the aim of the workshop was to receive input on the acceptability of its approaches to agricultural biotechnology. In response to criticisms about the narrow range of participants in previous consultation exercises, all interested parties were invited to attend, including consumer groups, environmentalists, and farmers, along with labour, animal welfare groups, and other interest groups. The opening speech by Agriculture Canada Assistant Deputy Minister Brian Morrissey set the tone for the workshop, however, stressing that with genetic modification, the "*potential benefits are virtually limitless*" (AAFC, HC, and EC 1993, 3; emphasis added). Morrissey proceeded to set the boundaries for discussion, informing the participants that the consultations would focus only on the scientific data requirements for assessing product safety. Thus, while the hosting departments were seeking a broader base for input that included nonscientists, the agenda was preset to limit discussion to scientific and technical issues.

The industry and researcher representatives at the workshop stressed the need for science-based regulations. They argued that products should be assessed on the basis of environmental and human health risks alone, and that there should not be any consideration of social and ethical issues. Industry officials continued to stress the need for the Canadian biotechnology industry to stay competitive and for regulations to be harmonized internationally. Environmentalists, represented by the Canadian Environmental Network, took advantage of the opportunity to raise their concerns with the new Federal Framework. In particular, they criticized the lack of transparency, accountability, public involvement, and access to information, argued that a public debate was required, and called for new legislation (AAFC, HC, and EC 1993). Although the government departments stated that these views were considered, there is no evidence that they had any impact. The boundaries imposed by Morrissey, restricting discussion to the scientific issues and framing biotechnology as beneficial, considerably marginalized those advocating consideration of the social and ethical issues.

The Emergence of Social and Ethical Concerns

Until the mid-1990s, social and ethical issues were largely absent from the debate leading to the establishment of the Federal Framework for Biotechnology, and they remained poorly understood by scientists, industry, and government. In addition, a limited number of restrictive public surveys and consultations had failed to engage the public and stakeholders on these issues (Mullen 1994; Industry, Science and Technology Canada 1993).[1] In the late 1990s, however, the growing public backlash in Europe highlighted the importance of public acceptance if commercialization of agricultural biotechnologies was to forge ahead. Demands for public involvement and consideration of the social and ethical issues came primarily from parliamentary reports and government bodies. Most of the recommendations were based on the premise that open consideration of these issues would increase public confidence and trust, which in turn would result in public acceptance.

In the late 1990s, three official reports addressed the social and ethical issues of agricultural biotechnology. The first was the House of Commons Standing Committee on the Environment and Sustainable Development report *Biotechnology Regulation in Canada: A Matter of Public Confidence* (1996). The committee recommended the establishment of a new, independent commission, with representatives from the public, government, industry, and the environmental, academic, and ethical communities, to consider the ethical aspects of biotechnology, particularly patenting, labelling, public trust in regulators, assessment of social benefits, and the flaws in risk-cost-benefit analysis (House of Commons Standing Committee on Environment and Sustainable Development 1996). In its official response, the government agreed on the need to consider the social and ethical issues, and envisaged these issues being considered in the broader policy framework, leaving the regulatory process a technical and scientific endeavour (Government of Canada 1999).

The second official report was the House of Commons Standing Committee on Agriculture and Agri-Food report *Capturing the Advantage: Agricultural Biotechnology in the New Millennium* (1998). This report was primarily concerned with capitalizing on the benefits of biotechnology, thereby furthering the goal of international competitiveness. The report advocated the "deficit model," where a risk communication strategy would educate an uninformed public about the benefits and risks that were to be predetermined by experts, in order to increase public acceptance.[2] Public concerns about social and ethical issues were considered a result of lack of scientific knowledge, so these concerns were marginalized.

The third official report to address social and ethical issues was the National Biotechnology Advisory Committee's *Leading into the Next Millennium* (1998). The report was supportive of the biotechnology industry, but for the first time NBAC recognized that public confidence was an issue that needed to be addressed if the biotechnology industry was to succeed in its goal of international competitiveness. In a shift away from the traditional deficit model of public communication, NBAC argued that two-way communication with the public and consideration of the social and ethical issues was required, and recommended that its role be expanded to take these issues into consideration (NBAC 1998).

The Federal Government Response to the Emergence of Social and Ethical Issues

The recommendations from parliamentary reports and government bodies noted above argued that considering the social and ethical issues would increase public confidence and therefore acceptance of GM crops. The government responded by establishing a committee with responsibility for social and ethical issues and requesting the Royal Society of Canada to report on the scientific issues related to human and animal health and environmental impacts.

Canadian Biotechnology Advisory Committee

In 1998, the government announced the new Canadian Biotechnology Strategy, which was intended to balance social and ethical issues with the primary goal of competitiveness (McDonald 2000). The strategy established the Canadian Biotechnology Advisory Committee in 1999 to advise on the policy-related ethical, social, regulatory, economic, environmental, and health aspects of biotechnology. CBAC reported to the Biotechnology Ministerial Coordinating Committee, which was coordinated by the minister of industry and consisted of the ministers of Agriculture and Agri-Food Canada, Health Canada, Environment Canada, Natural Resources Canada, Fisheries and Oceans, and Foreign Affairs and International Trade. A board of twenty-one external experts represented a fairly broad range of expertise from the fields of business, nutrition, law, faith, philosophy, ethics, and public advocacy (CBAC 2002a).

In 2000, CBAC held public consultations through multistakeholder workshops in five cities across the country to explore the social and ethical issues associated with regulating agricultural biotechnology, focusing on GM foods.[3] The workshops were expert consultations rather than public consultation exercises, in part because they lacked lay public and NGO involvement (Hartley and Skogstad 2005). For example, the Council of Canadians and eighty other NGO organizations refused to take part because of perceived bias in the mandate and process of the consultation, particularly the location of CBAC within Industry Canada and the ties between various CBAC members and industry (Hartley and Skogstad 2005).

CBAC's final report, *Improving the Regulation of Genetically Modified Foods and other Novel Foods in Canada,* was published in August 2002. Its main conclusions focused on the need for better-defined regulatory oversight, better research and data collection for monitoring the long-term health and environmental impacts, and a voluntary, rather than mandatory, labelling system. CBAC recommended that other laws and policies deal with the relevant social and ethical issues, and advised that the current regulatory framework should not be modified to incorporate these concerns due to the potential for abuse for political reasons, reduced predictability, and a commitment to international obligations (CBAC 2002b). The efficacy with which CBAC has managed to fulfill its mandate is questionable. Despite its mandate to advise on the policy-related ethical, social, regulatory, economic, environmental, and health aspects of biotechnology, CBAC demonstrates a limited understanding of the social and ethical issues and an inherent belief in the fact/value dichotomy. It envisioned insulating the scientific risk assessment from nonscientific issues, and argued that social and ethical issues be dealt with through the political process or in the marketplace.

Royal Society of Canada Expert Panel
Although CBAC had a mandate to consider the social and ethical issues related to agricultural biotechnology, it was the Royal Society of Canada that presented the first serious challenge to the marginalization of social and ethical issues. In 2001, a Royal Society Expert Panel, chaired by Conrad Brunk and Brian Ellis, published *Elements of Precaution: Recommendations for the Regulation of Food Biotechnology in Canada* (Royal Society of Canada 2001). The report was particularly critical of the regulatory process for agricultural biotechnology products, arguing that government regulators and officials needed to maintain a more neutral and objective perspective with respect to the public debate. The panel also critiqued the treatment of social and

ethical issues and the government's insistence that scientific risk assessment could be objective and value-free. The terms of reference given to the Royal Society by Health Canada, the Canadian Food Inspection Agency, and Environment Canada limited its investigation to the scientific issues related to human and animal health and environmental impacts. The panel found this mandate constraining and questioned the degree to which science could be objective and value-free, arguing that value judgments often influenced both the scientific and extrascientific domains of the policy framework. The panel argued that the scientific risks to health and the environment were only a small part of the debate over agricultural biotechnology, which included a host of social and ethical concerns that needed consideration. These concerns included the increasing corporate control of the food and agricultural sector, the move from family-run farms to huge agribusinesses, and the impact of agricultural biotechnology on farmers in the developing world as well as increased reliance of these farmers on multinational corporations from the overdeveloped world. It argued that such social and ethical concerns were central to the agricultural biotechnology debate, particularly from the public's perspective.

Rather ironically, the Royal Society Expert Panel report displays the most progressive perspective on social and ethical concerns in the Canadian debate over agricultural biotechnology, despite the panel's mandate to investigate only the scientific aspects of the regulatory framework. The report recognized and unpacked the social and ethical concerns, separating them into three categories: health and environmental risks, socio-economic risks, and philosophical/metaphysical risks. The panel questioned the fact/value dichotomy so entrenched in the risk decision-making model, arguing that ethics and values were not easily separated from science or scientific concerns. It also pointed out that socio-economic risks existed and needed further investigation, particularly social risks, which were excluded from the scientific dialogue and marginalized with other social and ethical concerns, and hence not explored further.

Insulating the Regulatory Framework from Social and Ethical Concerns: The Case of GM Wheat

Development of genetically modified wheat in Canada captured the attention of stakeholders and raised issues of the scope of the regulatory framework for products of such research. The debate over the role of social and ethical concerns in the regulation of agricultural biotechnology in Canada is perhaps best exemplified by the debate over the socio-economic risks of GM wheat.

This debate fleshed out the arguments both for and against the inclusion of social and ethical concerns in the policy and regulatory framework and demonstrated the prevailing belief in the fact/value dichotomy that justified the separation of objective science from subjective social and ethical concerns.

As noted earlier, a number of industry representatives and Canadian Biotechnology Advisory Committee members argued for the insulation of the regulatory process from consideration of the socio-economic risks associated with GM wheat. Describing social and ethical concerns as "emotional," a representative from the Grain Growers of Canada captured this perspective: "As soon as you allow emotion to step in, all good science goes out the window" (House of Commons Standing Committee on Agriculture and Agri-Food 2003c, 1140). The national spokesperson for the Office of Biotechnology at the Canadian Food Inspection Agency framed socio-economic and ethical concerns as "religious" and argued against a governmental role in considering them (Hartley 2005). She justified the exclusion of social and ethical concerns from the regulatory framework on the grounds that the social sciences were less able than the natural sciences to assess and predict, suggesting that bureaucrats would need "crystal balls."

The high-profile and contentious issue of regulatory scrutiny of GM wheat emerged after Monsanto Canada applied to the CFIA in December 2001 for the unconfined release of GM wheat for commercial production. The importance of wheat to the economy of western Canada, the nature of the wheat export market, and farmers' experiences with GM canola helped generate concern within the farming community over the socio-economic risks of allowing commercial production of GM wheat. However, these socio-economic risks could not be considered by the CFIA in its assessment of Monsanto's application. This presented a very serious and rather ironic dilemma for the government: the overarching goal of competitiveness that had driven the establishment of the regulatory framework was now threatened by the restrictive nature of the regulations. The government had to find a way to consider socio-economic concerns without weakening the notionally science-based regulatory framework. In March 2003, Agriculture Minister Lyle Vanclief suggested that the socio-economic risks would, indeed, be considered in the regulatory framework (Jeffs 2003).

The Canadian Wheat Export Market
At the time, wheat was a leading agricultural product in Canada. Most farmers in western Canada grew wheat as part of their crop rotation, generating gross revenue of approximately $3 billion annually (Wells and

Penfound 2003). Much of this revenue depended on an international customer base, with about 84 percent of domestic wheat production being exported (House of Commons Standing Committee on Agriculture and Agri-Food 2003c). Up to this point, farmers had generally been supportive of agricultural biotechnology in Canada. However, the fear of economic risks brought together farmers and environmentalists in opposition to GM wheat. Opponents included the powerful Canadian Wheat Board, the National Farmers Union, the Canadian Federation of Agriculture, the organic farming industry (represented by the Saskatchewan Organic Directorate), the Canadian Seed Growers' Association, Greenpeace, and the Council of Canadians.[4] Their concerns were based on mounting evidence that customers in Canada's large wheat export market would reject all Canadian wheat if Canada began commercial production of GM varieties of wheat, due to fears of contamination of non-GM products with GM material (Wells and Penfound 2003). For example, Rank Hovis, Britain's biggest flour mill, had stated publicly that if Canada grew GM wheat it would cease all Canadian wheat purchases, whether GM or conventional (Wells and Penfound 2003). In addition to these risks, many farming organizations argued that GM wheat offered no substantive benefits to farmers yet threatened potential (and unknown) increased costs for handling, storage, and shipping; liability; and unacceptable risks to the organic sector through contamination (Hall 2003).

Proponents of GM wheat, including Agriculture and Agri-Food Canada (AAFC), the Grain Growers of Canada, CropLife Canada, and the Western Canadian Wheat Growers Association, argued that contamination could be alleviated by the establishment of a segregation system, both in the field and in the supply chain. They further argued that segregation in the supply chain would not place an unfair cost burden on farmers. Others, however, did not believe that segregation was a viable solution to the potential problem of contamination. For example, W.H. Furtan, R.S. Gray, and J.J. Holzman (2003) argued that segregation was not possible; the commercial production of GM wheat would therefore result in lower market prices for all producers. The authors estimated that if foreign and domestic customers stopped buying Canadian wheat, Canadian farmers were likely to lose approximately $78.1 million (Furtan, Gray, and Holzman 2003).[5] Further, the inevitable resulting environmental contamination would lead to greater herbicide costs for all farmers, whether or not they adopted GM wheat.

The Impact of Socio-Economic Risks on the Regulatory Approval of GM Wheat

A high-profile debate took place in the public sphere, gaining significant media attention and focusing on the question of whether social and ethical concerns (in this case, the socio-economic risks) should and could be considered in risk decision making. Proponents wanted the regulatory framework to remain exclusively science-based and argued that social and ethical concerns should be left to the market to decide, believing that it was not the role of government to assess and determine market acceptance. This argument was based on the assumption that farmers who chose not to grow GM wheat could be protected against the risks of contamination, through either prevention or liability.

In addition, proponents of GM wheat argued that adding social and ethical concerns to the science-based regulatory framework would result in nontariff barriers to international trade, would be in conflict with international trade agreements, and would reduce Canada's ability to challenge countries that had established similar nontariff barriers to trade (House of Commons Standing Committee on Agriculture and Agri-Food 2003a).[6] Opponents argued that leaving it to the market to decide was problematic. The National Farmers Union cited two reasons. First, consumers did not have access to information through a labelling regime, and therefore would not be able to make an informed decision. Second, the issue of contamination meant that without a liability system in place, farmers who wanted to produce non-GM wheat might be harmed economically. At the time, no liability scheme was in place, although the very recent Supreme Court of Canada ruling in the case of *Monsanto Canada Inc. v. Schmeiser* indicated that liability might be the problem of the farmer and not the seed supplier.[7] Supported by local governments, 210 industry associations, 50 experts and researchers, 60 international organizations, and citizens' groups across Canada called for a moratorium on GM wheat until this issue and the issue of segregation could be dealt with (CBAC 2002a).

Opponents argued that the socio-economic risks of ignoring these concerns were great enough to warrant changes to the regulatory framework. The Canadian Wheat Board's solution, supported by the Canadian National Millers Association and the Saskatchewan Association of Rural Municipalities, among others, was to add a cost-benefit test to the existing regulatory approval process (House of Commons Standing Committee on Agriculture and Agri-Food 2003a). The test was envisaged to be industry-driven in that industry or stakeholders could request the cost-benefit analysis, but it would be a regulatory requirement, not a voluntary option. In this way, the cost-benefit test would be triggered only when there was concern about market acceptability, and would not be mandatory for all new varieties. The Canadian Federation of Agriculture also opposed GM wheat and wanted the existing science-based regulatory framework to remain as it was. However, it argued for marketing and segregation issues to be resolved through regulatory provisions before registration was granted (House of Commons Standing Committee on Agriculture and Agri-Food 2003d).

Consideration of Market Acceptability as Part of Regulatory Approval of GM Wheat

In April 2003, at the height of the debate, it became known (through an Access to Information request) that market acceptability had been used by the CFIA as a criterion during assessment for variety registration. The Prairie Registration Recommending Committee for Grain, which reviewed applications for variety registration, was informed by the CFIA in 2001 that the market acceptability criterion should no longer be used. A number of these types of recommending committees operated under the authority of the *Seeds Act*, overseeing the scientific assessment of new varieties. The committees recommended new varieties to the variety registration office at the CFIA, who then either accepted or rejected the recommendation. Registration was based on a number of criteria, including merit, agronomics, disease resistance, and milling quality. A new variety must show superior qualities to existing, registered seeds. Between 1990 and 2001, the criteria for evaluating new varieties included a "definition of merit" clause that allowed the committees to consider market acceptability (Wilson 2003). The government decided, however, that assessing market acceptability was not the role of government; rather, it should be left to industry to decide (CBC News 2003). The CFIA's vice president of operations, Peter Brackenridge, feared that the subcommittee might deny registration on the basis of social and ethical concerns, and therefore the criterion had to be removed, particularly as it had no statutory basis under the *Seeds Act* (House of Commons Standing Committee on Agriculture and Agri-Food 2003b).

The GM wheat issue was referred by Parliament to the House of Commons Standing Committee on Agriculture and Agri-Food, but although hearings took place for much of 2003, no report was generated. Apart from the agriculture minister's insistence that the problem of socio-economic risks would be addressed, the government did not formally respond to the debate on the GM wheat issue. CBAC, the advisory body responsible for extrascientific concerns, was not asked formally for comment, even though this issue was entirely within its mandate.

On May 10, 2004, Monsanto announced that it would defer its application for the environmental release of its Roundup Ready wheat, effectively bringing to a close the debate on GM wheat. The company's executive vice president, Carl Casale, noted: "As a result of our portfolio review and dialogue with wheat industry leaders, we recognize the business opportunities with Roundup Ready spring wheat are less attractive relative to Monsanto's other commercial priorities" (Monsanto Canada 2004, 1). Despite Monsanto's earlier claims that it would pursue commercial environmental release in spite of growing resistance, it seems likely that opposition to GM wheat directly influenced the company's decision to defer the application. Nevertheless, and despite Agriculture Minister Lyle Vanclief's statement to the contrary, the science-based regulatory framework remained insulated from social and ethical concerns.

Discussion

Overall, the social and ethical concerns over agricultural biotechnology in Canada have been characterized by experts as "tampering with nature" or "playing God," rather than as issues of social or environmental justice. Parliamentary committee debates have shed some light on the substance of these social and ethical concerns, including consumer choice, corporate control over the food and agricultural systems, impacts on small-scale farming communities, intellectual property rights, and shifting power relations between the farmer and the seed provider (House of Commons Standing Committee on Agriculture and Agri-Food 2003c).

For the most part, however, social and ethical concerns have been marginalized by the narrowly defined relationship between science and society that sees the public as a receiver of knowledge rather than having knowledge to contribute to policy and regulatory discussions. The absence of public consultation has meant that government officials lack understanding about these issues and have been quick to dismiss them. Although it is difficult to assess the concerns of the Canadian public, the limited available data on public perceptions appear to show that the public was far less concerned about these issues than publics in other countries, particularly Europe but also the United States. Indeed, it was stakeholders and experts – including

NGOs, farmers, and the Royal Society – who were chiefly responsible for advocating the inclusion of social and ethical concerns in the current framework. Environmental NGOs such as Greenpeace and Friends of the Earth were unsuccessful in getting social and ethical concerns to resonate with the public, and chose to focus instead on the health and safety risks of GM food (Hartley and Scott 2006).

Despite the general lack of understanding and marginalization of social and ethical issues in the Canadian regulatory context, one issue did infiltrate the risk decision-making model and became extremely influential in the debate surrounding agricultural biotechnology: international competitiveness. The government spent considerable sums emphasizing the benefits and playing down the risks of agricultural biotechnology to further this goal. For example, it invested over $13 million in public relations campaigns to promote the biotechnology industry (Aubry 2003). Greenpeace Canada and the Canadian Health Coalition accused the biotechnology industry of asking Agriculture and Agri-Food Canada for public money to counter claims that GM food was unsafe, and the government of paying over $3 million for information pamphlets insisting that GM food was safe (Freeze 2002). This project was extended globally with the Canadian government's request that Asia-Pacific Economic Cooperation (APEC) countries conduct similar education campaigns (Freeze 2002). In addition, the Canadian International Development Agency (CIDA) was accused of funding promotional campaigns for GM crops in developing countries; for example, it provided Monsanto with public money to encourage Chinese farmers to grow GM corn and cotton (Sharratt 2012).

Scientism has been described as a culture (Wynne 2006) and a discourse – a realm of ideas, concepts, categories, and beliefs that we take for granted (Kleinman 2005). Essentially, scientism describes the institutionalization of science within ideational and institutional frameworks in which policy making takes place. Central to scientism is the positivist dichotomy between facts and values, which allows science to be seen as objective and value-neutral. The authority of science is based precisely on this separation of facts and values, privileging facts over values. Scientism structures the relationship between science and society in a way that privileges scientific experts over the lay public, as experts have the knowledge and ability to identify the facts. If science is objective, it has a monopoly on the truth, as the only source of rational enquiry and decision making. In this privileged position, scientists become arbiters of controversy (Kleinman and Kinchy 2003).

In the case of agricultural biotechnology, including agricultural genomics, the fact/value dichotomy makes science a popular policy tool because it provides a neutral and parsimonious framework that offers clarity and fairness to product developers. As values diverge to a greater degree than facts, values reduce consensus and leave decisions open to attack. However, critics of positivist science have long argued that facts and values are not easily separated, particularly in regulatory decision making, where risk is concerned (Meyer 2011). Risk decision making is characterized by uncertainty, unknowns, complexity, missing or inadequate data, time constraints, and political pressures that science does not acknowledge (Brunk, Haworth, and Lee 1991). Assumptions made during risk decision making are not simply a result of the problems of "regulatory" science. Critics argue that even in laboratory situations, scientific methodologies require decisions and assumptions to be made that involve value judgments, and in this way influence outcomes (Shrader-Frechette 1985).

Scientism can shape the degree to which social and ethical concerns are considered by determining their legitimacy and the authority of those who champion them in several ways. First, scientism characterizes social and ethical concerns as "public" concerns that are based on ignorance and/or emotional responses. Wynne (2006) has argued that culturally entrenched scientism results in public issues involving science being framed as "scientific issues," and that if the public raises concerns about such "scientific issues," they are characterized as mistrustful of science or as anti-science. These assumptions privilege experts and support the "deficit model," where publics are considered ignorant and emotional and in need of education about the scientific facts in order to bring their "perceptions" of the issues in line with those of the scientific experts (McHughen 2000; Isaac 2006).

Second, all concerns can be framed as scientific concerns. Opponents to a type of technology can gain entry to the debate only if their concerns focus on the human health or environmental safety issues or are framed as "scientific" (Kleinman and Kinchy 2003; Hartley and Scott 2006). Scientism privileges human, health, and environmental safety over social and ethical concerns, as human, health, and environmental safety risks can be assessed by science, whereas social and ethical issues are determined by value judgments. Scientism as a discourse controls what issues can be legitimately discussed, marginalizing value-based concerns (Kleinman and Kinchy 2003). Finally, scientism relegates social and ethical issues to the realm of value concerns and, while important, such value concerns should be considered in isolation from science.

This historical analysis demonstrates that despite opportunities to consider the social and ethical issues, the scientific determination of risk has continued to dominate the policy and regulatory framework with respect to agricultural biotechnology, and the social and ethical issues have remained largely excluded from the debate. The Canadian government has adopted a science-based regulatory framework for the case-by-case assessment of individual products. With the exception of the government's goal of international competitiveness, social and ethical issues played a marginal role in the development of this regulatory framework and the broader policy framework.

Since 2005, there have been no significant changes to the regulation of agricultural biotechnology or genomics in Canada, and no attempts have been made by government to explicitly consider the social and ethical concerns raised by such technologies. In 2007, the government announced the termination of CBAC in favour of a new advisory committee that would be more in line with its goal of international competitiveness (Government of Canada 2007). CBAC's demise came as no surprise. The committee's inherent problems grew increasingly obvious to its members and the external community, including government, and as time passed there was little evidence that the government took its work seriously (CBAC 2004). As of 2014, the regulatory framework continues to be focused solely on the scientific safety of commercial products. As the Canadian Food Inspection Agency has stated, "the issue of whether or not these products are 'necessary' is left to the market place to determine" (CFIA 2014).

The regulatory framework described in this chapter applies to all products of agricultural biotechnology, including GM crops and products from the next-generation "omics" technologies, such as genomics and proteomics. Agricultural genomics research, in turn, may contribute to the development of a variety of products not necessarily limited to those developed through genetic modification. The novelty of such products may trigger the application of a wide spectrum of regulatory regimes directed at novel products (for further discussion, see Chapter 1). This chapter demonstrates that the debate over social and ethical issues emerged after the regulatory framework for agricultural biotechnology was established in the early 1990s. Although the science-based framework faced several challenges, such as the case of GM wheat, it has remained largely insulated from social and ethical issues. At present, emerging agricultural genomics products will be evaluated through this science-based risk assessment, with no forum for considering their social and ethical impacts.

In recent years, the concept of Responsible Research and Innovation (RRI) has emerged as a popular and important innovation governance concept in Europe (Macnaghten et al. 2014; Owen, Bessant, and Heintz 2013). RRI attempts to increase the public value of innovation through the inclusion of the public and stakeholders upstream in the innovation process. The social and ethical issues associated with possible innovation trajectories are collectively anticipated and reflected on, and these issues are able to steer the innovation processes and outcomes to more socially desirable ends. Essentially, the European Union and its member states are employing RRI to better align research and innovation with societal needs (European Commission 2013). For example, RRI forms a significant pillar of the EU Framework Programme for Research and Innovation, Horizon 2020, and has been institutionalized within research funding institutions in the United Kingdom, the Netherlands, and Norway (von Schomberg 2013).

Agricultural biotechnology is continuing to develop products from large-scale investments in agricultural genomics. Although genomics has not, to date, generated public or stakeholder concern on the order seen for GM technologies, it has the potential to introduce novel plants and new traits. Ultimately, however, based on the historical developments outlined in this chapter, any assessment of risks posed by these genomics products will be conducted through the existing science-based risk assessment, with no forum for considering the social and ethical issues.

In the context of this volume's larger discussion of the Intellectual Property–Regulatory Complex and its ultimate impact on the development of socially beneficial products of agricultural genomics, the regulatory system outlined here operates on a national level as the main thrust of Canada's response to all innovative plant products. Its overt focus on scientific risks to the exclusion of most socio-economic considerations reflects the priorities of such international accords as the *General Agreement on Tariffs and Trade* (GATT) and the *Agreement on the Application of Sanitary and Phytosanitary Measures* (SPS Agreement). Similar priorities are shared by some potential importers of such products and many of Canada's larger trading partners. As they move into other jurisdictions, however, developers must also be aware that priorities and policy choices can vary widely, and many other countries (including many developing countries) tend to emphasize social considerations and community perspectives to a much greater extent.[8]

These differences, especially Canada's explicit exclusion of social considerations and perspectives that are critically important for many nations, exemplify the difficulties posed by the IP–Regulatory Complex as it relates to the delivery of products and translation of research in this space. The concept of RRI may have utility in the Canadian IP–Regulatory Complex, allowing collective and inclusive upstream debate on the social and ethical dimensions of agricultural genomics innovations to improve the translation of publicly funded research into socially beneficial products.

Notes

1 In 1993, Industry, Science and Technology Canada (1993) held a round of focus groups in order to better market biotechnology to the public. The exercise found that biotechnology was poorly understood and that the public may perceive biotechnology negatively. In the same year, the first public opinion survey on biotechnology conducted in Canada revealed that public confidence in the biotechnology industry and government regulators was moderately low (Mullen 1994).
2 See Wynne 2006 for a discussion of the deficit model as an approach to the science/society relationship.
3 CBAC commissioned a series of research reports between 1999 and 2001 to investigate explicitly the broader issues laid out in its remit. Most of these reports deal directly with social and ethical issues of biotechnology. CBAC had the freedom to choose the topics for reports. See Thompson 2000 and Silverman 2000 for examples of these reports.
4 The Canadian government removed the Wheat Board's monopoly over wheat and barley sales in 2012, considerably reducing its power. See *Marketing Freedom for Grain Farmers Act*, S.C. 2011, c. 25.
5 Monsanto would be set to make about $157 million (Hanley 2003).
6 Canada had joined the United States in a World Trade Organization challenge against the European Union, claiming that the latter's moratorium on GM products represented a nontariff barrier to trade.
7 On May 21, 2004, the Supreme Court ruled 5–4 that Percy Schmeiser had violated Monsanto's patent on Roundup Ready canola seeds. Schmeiser had argued that Monsanto's GM seed had cross-pollinated with his own seed, leading to contamination. Monsanto argued that Schmeiser had stolen the seed, infringing its patent rights. The case was portrayed in the media as a "David-versus-Goliath" story, pitting the US multinational, backed by subsidies from the Canadian government, against the small prairie farmer (Beattie 2004). Commenting on the ruling, CBAC argued that the decision was heavily weighted in favour of industry to the detriment of farmers (Gold 2004). A more recent case brought to the Saskatchewan court by the Saskatchewan Organic Directorate in 2004 addressed the issue of liability more directly. The Saskatchewan Organic Directorate complained that organic farmers had suffered economic hardship due to the contamination of organic crops by GM canola, and had stopped growing organic canola due to the high risk of contamination. As a result, the Organic Directorate sought class certification to sue Monsanto and Bayer for such losses. However, the judge ruled that the farmers

had not demonstrated economic hardship, and rejected the application for class certification. See *Hoffman v. Monsanto Canada,* 2005 SKQB 225.

8 In any event, as made clear in Chapter 6, socio-economic factors, whether explicitly considered or not, often play a key role in shaping the market and other incentives available for developers in many nations.

References

Legislation
Feeds Act, R.S.C., 1985, c. F-9.
Health of Animals Act, S.C. 1990, c. 21.
Marketing Freedom for Grain Farmers Act, S.C. 2011, c. 25.
Pest Control Products Act, S.C. 2002, c. 28.
Seeds Act, R.S.C., 1985, c. S-8.

Cases
Hoffman v. Monsanto Canada Inc., 2005 SKQB 225.
Monsanto Canada Inc. v. Schmeiser, [2004] 1 S.C.R. 902, 2004 SCC 34.

Secondary Sources
AAFC (Agriculture and Agri-Food Canada), HC (Health Canada), and EC (Environment Canada). 1993. *Workshop on Regulating Agricultural Products of Biotechnology, Nov 8–10: Proceedings.* Ottawa: Agriculture and Agri-Food Canada.

Aubry, Jack. 2003. "13M Polishes Biotech Image, Critics Charge: Health Advocates Say Federal Funds Should Go to Testing, Labelling GM Foods." *Ottawa Citizen,* November 24, A5.

Barrett, Katherine J. 1999. "Canadian Agricultural Biotechnology: Risk Assessment and the Precautionary Principle." PhD dissertation, University of British Columbia.

Beattie, Adrienne. 2004. "No Justice." *Calgary Herald,* May 27, A16.

Brunk, Conrad G., and Sarah Hartley, eds. 2012. *Designer Animals: Mapping the Issues in Animal Biotechnology.* Toronto: University of Toronto Press.

Brunk, Conrad, Lawrence Haworth, and Brenda Lee. 1991. *Value Assumptions in Risk Assessment: A Case Study of the Alachlor Controversy.* Waterloo, ON: Wilfrid Laurier University Press.

Bull, Alan T., Geoffrey Holt, and Malcolm D. Lilly. 1982. *Biotechnology: International Trends and Perspectives.* Paris: OECD.

Bunton, Robin, and Alan Peterson, eds. 2005. *Genetic Governance: Health Risk and Ethics in the Biotech Era.* New York: Routledge.

Caldwell, R.B., and L.H. Duke. 1988. *The Regulation of Plant Biotechnology in Canada.* Ottawa: Agriculture Canada, Seeds Division.

Canadian Agricultural Research Council. 1988. *Proceedings of the Workshop on Regulation of Agricultural Products of Biotechnology.* Ottawa: Queen's Printer.

CBAC (Canadian Biotechnology Advisory Committee). 2002a. *2000–2001 Annual Report.* Ottawa: CBAC.

–. 2002b. *Improving the Regulation of Genetically Modified Food and Other Novel Foods in Canada: Report to the Government of Canada Biotechnology Ministerial Coordinating Committee.* Ottawa: CBAC.

–. 2004. "Completing the Biotechnology Regulatory Framework." Advisory memorandum. Ottawa: CBAC.

CBC News. 2003. "U.S. Producers Benefit from Canadian Beef Ban: Statistics Canada." CBC News, December 4. http://www.cbc.ca/news/business/u-s-producers-benefit-from-canadian-beef-ban-statistics-canada-1.384999.

CFIA (Canadian Food Inspection Agency). 2003. *Status of Regulated Plants with Novel Traits in Canada: Unconfined Environmental Release, Novel Livestock Feed Use, Variety Registration and Novel Food Use.* Ottawa: CFIA, Plant Biosafety Office, Plant Products Directorate.

–. 2014. "Regulating Agricultural Biotechnology in Canada: An Overview." CFIA, http://www.inspection.gc.ca/plants/plants-with-novel-traits/general-public/fact-sheets/overview/eng/1338187581090/1338188593891.

CIELAP (Canadian Institute for Environmental Law and Policy). 1988. *Biotechnology Policy Development. Volume II. Report Prepared for the Ontario Ministry of the Environment, No.265RR.* Toronto: CIELAP.

Dowdeswell, Elizabeth, Abdallah S. Daar, and Peter Singer. 2005. "Getting Governance into Genomics." *Science and Public Policy* 32 (6): 497–98. http://dx.doi.org/10.3152/147154305781779263.

European Commission. 2013. *Options for Strengthening Responsible Research and Innovation. Report of the Expert Group on the State of Art in Europe on Responsible Research and Innovation.* Luxembourg: Publications Office of the European Union. http://ec.europa.eu/research/science-society/document_library/pdf_06/options-for-strengthening_en.pdf.

Federal Ethics Committee on Non-Human Biotechnology. 2012. *Release of Genetically Modified Plants – Ethical Requirements.* Bern: Federal Ethics Committee on Non-Human Biotechnology.

Freeze, Colin. 2002. "Ottawa Promoting Safety of GMOs: Critics Say Money Spent on Food Pamphlets Shows Watchdog Too Close to Biotech Firms." *Globe and Mail,* February 5, A9.

Furtan, W.H., R.S. Gray, and J.J. Holzman. 2003. "The Optimal Time to License a Biotech 'Lemon.'" *Contemporary Economic Policy* 21 (4): 433–44. http://dx.doi.org/10.1093/cep/byg023.

Gold, E. Richard. 2004. "Monsanto's Gain Is Everyone Else's Pain." *Globe and Mail,* May 24, A17.

Gottweis, Herbert. 2005. "Emerging Forms of Governance in Genomics and Post-Genomics." In *Genetic Governance: Health Risk and Ethics in the Biotech Era,* ed. Robin Bunton and Alan Peterson, 175–93. New York: Routledge.

Government of Canada. 1999. *Government Response to the Third Report of the Standing Committee on the Environment and Sustainable Development Entitled Biotechnology Regulation in Canada: A Matter of Public Confidence.* Ottawa: Government of Canada.

–. 2007. "Science and Technology Strategy: Mobilizing Science and Technology to Canada's Advantage." Industry Canada, http://www.ic.gc.ca/eic/site/icgc.nsf/eng/h_00856.html.

Hall, Angela. 2003. "Genetically Modified Crop Worries Wheat Board." *StarPhoenix* (Saskatoon), January 9, C4.

Hanley, Paul. 2003. "GM Wheat a Real 'Lemon.'" *StarPhoenix* (Saskatoon), March 25, C3.

Hartley, Sarah. 2005. "The Risk Society and Policy Responses to Environmental Risk: A Comparison of Risk Decision-Making for GM Crops in Canada and the UK, 1973–2004." PhD dissertation, University of Toronto.

Hartley, Sarah, and Dayna Nadine Scott. 2006. "Out-of-Bounds? Resisting Discursive Limits in the Debate over Food Biotechnology." *Canadian Review of Social Policy* 57: 104–9.

Hartley, Sarah, and Grace Skogstad. 2005. "Regulating Genetically Modified Crops and Foods in Canada and the United Kingdom: Democratizing Risk Regulation." *Canadian Public Administration* 48 (3): 305–27. http://dx.doi.org/10.1111/j.1754-7121.2005.tb00228.x.

Henley, Doreen. 1986. *Co-ordinated Study on Government Processes in Safety and Regulation of Modern Biotechnology: Report to the Interdepartmental Committee on Biotechnology.* Ottawa: Interdepartmental Committee on Biotechnology.

House of Commons Standing Committee on Agriculture and Agri-Food. 1998. *Capturing the Advantage: Agricultural Biotechnology in the New Millennium.* Third Report. Ottawa: Public Works and Government Services. Parliament of Canada, http://www.parl.gc.ca/HousePublications/Publication.aspx?DocId=1031520&Language=E&Mode=1&Parl=36&Ses=1&File=6.

–. 2003a. *Evidence.* 37th Parliament, 2nd Session, April 3, 1110–1250. Parliament of Canada, http://www.parl.gc.ca/HousePublications/Publication.aspx?DocId=820530&Language=E&Mode=1&Parl=37&Ses=2.

–. 2003b. *Evidence.* 37th Parliament, 2nd Session, May 15, 1115–1255. Parliament of Canada, http://www.parl.gc.ca/HousePublications/Publication.aspx?DocId=917370&Language=E&Mode=1&Parl=37&Ses=2.

–. 2003c. *Evidence.* 37th Parliament, 2nd Session, June 5, 1115–1300. Parliament of Canada, http://www.parl.gc.ca/HousePublications/Publication.aspx?DocId=968241&Language=E&Mode=1&Parl=37&Ses=2.

–. 2003d. *Evidence.* 37th Parliament, 2nd Session, June 12, 1110–1310. Parliament of Canada, http://www.parl.gc.ca/HousePublications/Publication.aspx?DocId=1009747&Language=E&Mode=1&Parl=37&Ses=2.

House of Commons Standing Committee on Environment and Sustainable Development. 1996. *Biotechnology Regulation in Canada: A Matter of Public Confidence.* Third Report. Ottawa: Parliament of Canada.

Howlett, Michael, and David Laycock, eds. 2012. *Regulating Next Generation Agri-Food Biotechnologies: Lessons from European, North American and Asian Experience.* Abingdon, UK: Routledge.

Industry, Science and Technology Canada. 1993. *Attitudes of Canadians towards Biotechnology: Four Focus Groups.* Final Report. Ottawa: ISTC.

Isaac, Grant E. 2006. "Biotechnology Policy and the Risk Analysis Framework." In *Governing Risk in the 21st Century: Lessons from the World of Biotechnology,* ed. Peter W. B. Phillips, Stuart Smyth, and William A. Kerr, 31–43. New York: Nova Science.

Jeffs, Allyson. 2003. "Wheat Sales Won't Be Risked: Impact on Exports Must Be Studied – Agriculture Minister." *Edmonton Journal,* March 13, H1.

Kalous, M.J., and L.H. Duke. 1989. *The Regulation of Plant Biotechnology in Canada: Part 2, The Environmental Release of Genetically Altered Plant Material.* Ottawa: Agriculture Canada, Seeds Division.

Kleinman, Daniel Lee. 2005. *Science and Technology in Society: From Biotechnology to the Internet.* Malden, MA: Blackwell Publishing.

Kleinman, Daniel Lee, and Abby J. Kinchy. 2003. "Why Ban Bovine Growth Hormone? Science, Social Welfare, and the Divergent Biotech Policy Landscapes in Europe and the United States." *Science as Culture* 12 (3): 375–414. http://dx. doi.org/10.1080/09505430309010.

Kneen, Brewster. 1992. *The Rape of Canola.* Toronto: NC Press.

Laycock, David, and Michael Howlett. 2012. "Regulating Next Generation Bio-technologies: Tentative Regulation for Emerging Technologies." In *Regulating Next Generation Agri-Food Biotechnologies: Lessons from European, North American and Asian Experiences,* ed. Michael Howlett and David Laycock, 3–12. Abingdon, UK: Routledge.

Macnaghten, P., R. Owen, J. Stilgoe, et al. 2014. "Responsible Innovation across Borders: Tensions, Paradoxes and Possibilities." *Journal of Responsible Innovation* 1 (2): 191–99. http://dx.doi.org/10.1080/23299460.2014.922249.

McDonald, Michael. 2000. *Biotechnology, Ethics and Government: A Synthesis.* Prepared for the Canadian Biotechnology Advisory Committee, Project Steering Committee on Incorporating Social and Ethical Considerations into Biotech-nology. Ottawa: Canadian Biotechnology Advisory Committee.

McHughen, Alan. 2000. *Pandora's Picnic Basket: The Potential and Hazards of Genetically Modified Foods.* New York: Oxford University Press.

McIntyre, T.C. 1990. *Asleep at the Switch: The Federal Government, Environment, and Planning for High Technology. The National Biotechnology Strategy, 1983– 1993.* PhD dissertation, University of Waterloo.

Meyer, Hartmut. 2011. "Systemic Risks of Genetically Modified Crops: The Need for New Approaches to Risk Assessment." *Environmental Sciences Europe* 23 (7). http://dx.doi.org/10.1186/2190-4715-23-7.

Monsanto Canada. 2004. "Monsanto to Realign Research Portfolio, Development of Roundup Ready Wheat Deferred: Decision Follows Portfolio Review, Consulta-tion with Growers." Press release, May 10.

Mullen, Michelle. 1994. *Biotechnology: Social and Ethical Issues: Industry's Com-mitment, and Public Policy.* Toronto: Ontario's Biotechnology Advisory Board.

NBAC (National Biotechnology Advisory Committee). 1988. *Annual Report 1987– 1988. The Regulation of Biotechnology: A Critical Issue for Canadian Research and Industrial Development.* Ottawa: Minister of Supply and Services Canada.

–. 1998. *Leading into the Next Millennium.* Sixth Report. Ottawa: Industry Canada.

Nuffield Council on Bioethics. 2012. *Emerging Biotechnologies: Technology, Choice and the Public Good.* London: Nuffield Council on Bioethics.

Owen, Richard, John Bessant, and Maggy Heintz, eds. 2013. *Responsible Innovation: Managing the Responsible Emergence of Science and Innovation in Society.* Chichester, UK: John Wiley and Sons. http://dx.doi.org/10.1002/9781118551424.

Phillips, P.W.B., D. Castle, S.J. Smyth, et al. 2010. "A Response to the Nuffield Council on Bioethics Consultation Paper "New Approaches to Biofuels." Valgen, http://valgen.ca/wp-content/uploads/2010/04/A-Response-to-the-Nuffield-Council-on-Bioethics-Consultation-Paper-final.pdf.

Royal Society of Canada. 2001. *Elements of Precaution: Recommendations for the Regulation of Food Biotechnology in Canada. Expert Panel Report on the Future of Food Biotechnology.* Ottawa: Royal Society of Canada.

Scowcroft, W.R., and L. Holbrook. 1988. *Regulation of Plant Biotechnology – Perception, Reality and Risk. Canadian Agricultural Research Council Workshop, December 5–6: Regulation of Agricultural Products of Agricultural Biotechnology.* Ottawa: Canadian Agricultural Research Council.

Sharratt, Lucy. 2012. "Genetic Engineering in Canada and the Right to Food Internationally." Canadian Biotechnology Action Network, http://www.cban.ca/content/download/1404/8869/file/CBAN%20Remarks%20to%20UN%20Rapporteur%20on%20the%20Right%20to%20Food.pdf.

Shrader-Frechette, K.S. 1985. *Science Policy, Ethics, and Economic Methodology: Some Problems of Technology Assessment and Environmental-Impact Analysis.* Dordrecht, Netherlands: D. Reidel.

Silverman, Ozzie. 2000. *International Approaches to Non-Science Issues in Regulating the Products of Biotechnology.* Prepared for the Canadian Biotechnology Advisory Committee Project Steering Committee on the Regulation of Genetically Modified Foods. Ottawa: Canadian Biotechnology Advisory Committee.

Thompson, Paul. 2000. *Food and Agricultural Biotechnology: Incorporating Ethical Considerations.* Prepared for the Canadian Biotechnology Advisory Committee Project Steering Committee on the Regulation of Genetically Modified Foods. CGIAR, https://research.cip.cgiar.org/confluence/download/attachments/3434/AO9.pdf.

von Schomberg, R. 2013. "A Vision of Responsible Research and Innovation." In *Responsible Innovation: Managing the Responsible Emergence of Science and Innovation in Society,* ed. Richard Owen, John Bessant, and Maggy Heintz, 51–74. Chichester, UK: John Wiley and Sons. http://dx.doi.org/10.1002/9781118551424.ch3.

Wells, Stewart, and Holly Penfound. 2003. "The Risks of Modified Wheat." *Toronto Star,* February 25, A26.

Wilson, Barry. 2003. "Market 'Risk' Once Part of Process." *Western Producer,* April 11. Canadian Health Coalition, http://www.producer.com/2003/04/market-risk-once-part-of-process/.

Wynne, Brian. 2006. "Public Engagement as a Means of Restoring Public Trust in Science – Hitting the Notes but Missing the Music?" *Community Genetics* 9 (3): 211–20. http://dx.doi.org/10.1159/000092659.

3

How the IP–Regulatory Complex Affects Incentives to Develop Socially Beneficial Products from Agricultural Genomics

GREGORY GRAFF AND DAVID ZILBERMAN

A central question raised in this volume is how regulatory and intellectual property (IP) systems affect incentives for the development of socially beneficial agricultural products derived from agricultural genomics research. To address this question, it will be helpful to establish first what we mean by "socially beneficial," and to compare common public perceptions of socially beneficial agricultural products with economic measures of the social costs and benefits associated with today's major commercial products derived from agricultural genomics. Next, we need to consider the effects of IP and regulatory systems on incentives to innovate in agriculture, first treating each independently and then bringing the two together into what we refer to in this book as the "IP–Regulatory Complex" in order to consider interactions between these two systems. We then lay out some of the impacts that these systems appear to be having on the rate and composition of innovation in agriculture. This chapter explores one of the main theses of this volume – namely, that IP and regulatory systems are intimately intertwined. Although in practice companies may maintain separate patent divisions and regulatory affairs divisions, the two areas affect each other intrinsically. Similarly, at the policy level, we have certain agencies administering intellectual property and others dealing with biosafety and environmental regulation separately, but, in fact, their decisions and actions interact intrinsically and affect economic activity and social benefit. Unsurprisingly, we designate ourselves as either IP experts or as regulatory experts, especially in academia, and are uncomfortable when we stray from our particular domain of expertise; inevitably, however, we find ourselves interacting with one another and drawing on one another's work.

Ultimately, this chapter is concerned with the kinds of incentives for innovation created by the IP–Regulatory Complex. The work is informed throughout by empirical glimpses into how the IP–Regulatory Complex may in fact be affecting innovation in product areas that meet our working definition of "socially beneficial." The chapter concludes with recommendations for improving incentives and facilitating the innovation of socially beneficial products.

Deriving Social Benefit from New Products in Agriculture

There is no set definition of "social benefit" with respect to agricultural products. Indeed, considering the vast array of products offered on the market, it can be quite difficult for consumers to evaluate systematically the social benefits and costs associated with the production and consumption of individual products. This is not surprising. After all, economists differ in their opinions on how to evaluate the impacts of market activities on economic equity (Just, Hueth, and Schmitz 2008) and engineers debate the proper scope of methods such as life-cycle analysis (LCA) used to evaluate the environmental footprint resulting from the "cradle-to-grave" production and utilization of a given product.[1] Consequently, consumers are largely influenced by retailers' marketing departments and the media's often sensationalized interpretations of science and industry. Under these influences, consumers often develop simple rules of thumb, criteria they believe to be correlated with the maximization of social or environmental benefits, or at least with the minimization of social or environmental harms. Examples of characteristics of agricultural products for which consumers have developed rules of thumb include the presence of recognized "good" ingredients (vitamins, antioxidants, omega-3 fatty acids, fibre, and so on), the absence of recognized "bad" ingredients (refined sugars, saturated fats, trans fats, and so on), the content of recycled materials, the reduction or elimination of packaging, and even the display of short stories and vignettes on the label giving information about the source and origins of the product (Costanigro et al. 2014; Heiman and Lowengart 2008). Consumers' selection criteria also include brandlike "quality marks" that signal adherence to standards that embody social conceptions of what is socially or environmentally beneficial. Examples of such quality marks used on agricultural products include "organic," "local," "free-trade," "wildlife-friendly," or "GMO-free" (Costanigro et al. 2011, 2014). Other marketing scholars point to "halo effects," where the presence of such cues influences consumers' evaluation of other attributes of the product, effectively biasing their subsequent perceptions and judgments in a positive way. For example, health claims on a food product label have been shown to induce consumers to express positive valuations of that product and to truncate their search early when shopping (Roe, Levy, and Derby 1999).

Yet – and this is crucial – the actual, objective qualities underlying such product label claims can vary widely, both in definition and in their actual impact on the environment and public health. First, the stringency or verifiability of compliance with the associated standards can vary widely. More important, however, the display of such characteristics or quality marks may be only roughly correlated with the maximization of personal health as well as social or environmental benefits.[2] Finally, the amount of benefit can depend greatly on the consumer's context, such as health conditions or where they live.

Not all consumers are equally responsive to such marketing, but consumers do, by and large, reveal in their purchasing behaviours a general preference for products perceived to be socially and environmentally beneficial, and some consumers clearly are willing to pay a premium on products that display such characteristics or quality marks (Costanigro et al. 2014; Seufert, Ramankutty, and Foley 2012).

Scientific inquiry, however, seeks more objective definitions and measures of social benefit. In economics, we turn to the methodology of social welfare analysis (Just, Hueth, and Schmitz 2008), looking at indicators such as production yields, price trends, and environmental and related public health data as indicators of the impact that the production and consumption of a product has on social welfare or environmental quality. Such analytical tools can be more reliable than the criteria used by consumers to ascertain the social benefit associated with products that result from agricultural genomics.

Understanding the Social Costs and Benefits of Today's Major Commercial Products from Agricultural Genomics

Agricultural genomics is a set of tools and a source of greater knowledge, capable of giving rise to a wide range of applications and product innovations in contemporary agriculture. It has given us an understanding of the genome and an identification of particular genes in major crop species, livestock species, and beneficial microbes as well as pathogens. This broad base of knowledge is resulting in a wide range of technical applications that enable the development of new products. The methods include identification of the mechanisms of crucial traits, diagnostics or detection of plant and animal disease or microbial contaminants, and genetic engineering and molecular breeding programs for the expression of new traits. Most such methods continue to further already-established trajectories of technological advance, in such areas as agricultural genetics, breeding, and pathology. However, the genetic engineering of traits into productive agricultural species in particular has introduced a fundamentally new trajectory of technological advance, challenging established modes of agricultural production and heralding a whole new range of possible applications of agricultural genomics.

To date, there are really only two major commercial uses of crop genetic engineering that have been introduced in agriculture: herbicide tolerance and insect resistance. Both of these uses are intended for on-farm pest control. For the end consumer, they do not change the resulting product in any readily detectable way.

The genetic trait of herbicide tolerance has been widely commercialized in soybean, maize, and canola, particularly in the United States, Canada, Brazil, and Argentina. The most common herbicide tolerance trait is due to the insertion of a gene encoding a modified enzyme that makes the plant resistant to the plant-killing action of the herbicide glyphosate (widely sold under the trademark "Roundup"). The fact that the crop can withstand the herbicide while nearby weeds cannot allows for a more efficient system of weed control to be employed on the farm, by spraying herbicide rather than mechanically tilling the soil. The resulting system of weed control by herbicide has led to reduced soil erosion and increased soil retention of organic matter and carbon, as well as water. A secondary effect has been a systemic shift toward the use of glyphosate and away from harsher herbicides.[3] Concerns have arisen, however, that the widespread use of glyphosate has resulted in the emergence of weeds that are themselves resistant to the herbicide, potentially undermining the efficacy of glyphosate and the benefits enjoyed by all farmers who depend on it for weed control (National Research Council 2010).

A genetic trait for insect resistance has been widely commercialized in maize and cotton, particularly in the United States, Canada, India, and China. This trait enables more efficient control of specific insect pests and replaces the need to spray chemical pesticides. These engineered crops produce small amounts of the *Bt* microbe's endotoxin proteins within their tissues.[4] These proteins are effective and highly targeted biopesticides that work against chewing and boring larvae of various lepidopteran (butterflies and moths) and coleopteran (beetles) species, such as corn rootworm, European corn borer, and pink bollworm. These are among the most damaging pests of corn and cotton globally, capable of causing billions of dollars' worth of crop damage in a given year and typically requiring application of billions of dollars' worth of chemical pesticides to combat (National Research Council 2010).

Although these genetic technologies have had massive impacts on agriculture, they tend not to be viewed positively by consumers. Moreover, a variety of activist nongovernmental organizations have protested against the commercial use of the products of genetic engineering. There is a widespread perception that these technologies have enriched oligopolistic corporations that control the technology through patents, trademarks, and other forms of intellectual property while harming the interests of consumers, small farmers, and the environment.[5] Some food-manufacturing companies voluntarily label their foods "GM-free," which is used as a kind of quality mark.[6] As a result of these controversies and mixed reviews, consumers are hesitant.[7] A recent survey of consumers in Colorado found that about 40 percent of respondents express confidence in genetically engineered foods, about 40 percent express concerns, and the remaining 20 percent feel they do not know enough about the technology to form an opinion.[8]

Social Welfare Analysis of *Bt* Corn

Objective measures of the impacts of these major commercial products on social welfare and the environment are far more positive than the public reception would suggest. According to one analysis of markets for farm commodity inputs and outputs, the impact of *Bt* insect-resistant corn on the main stakeholders in the US market illustrates how – as with virtually any economic policy or technology change – there are both winners and losers as a result of the changes ushered in with the new technology (Wu 2004). Overall, the use of the *Bt* trait in corn has led to economic benefits to the US economy that, according to this study, approached $1 billion a year in the early 2000s.

The greatest beneficiaries are consumers, who realized an estimated $520 million per year of benefit in the form of reduced prices for consumer products (see Figure 3.1). Such reduced prices are not limited to products that are recognizably made of corn. Consumers benefit from

Figure 3.1

The apportionment, among different stakeholder groups in the US corn seed industry, of benefits and costs resulting from the adoption of *Bt* corn technology

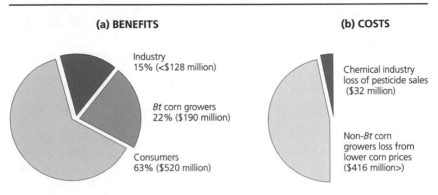

(a) BENEFITS

Industry
15% (<$128 million)

Bt corn growers
22% ($190 million)

Consumers
63% ($520 million)

(b) COSTS

Chemical industry
loss of pesticide sales
($32 million)

Non-*Bt* corn
growers loss from
lower corn prices
($416 million>)

Source: Adapted from Wu 2004.

follow-on indirect effects of lower-cost corn as well, including lower meat and dairy prices due to lower feed costs for livestock and dairy operations; lower prices of products containing sweeteners due to downward pressure on the price of corn syrup, which makes up a major share of the sweetener market; and somewhat lower gasoline prices due to lower costs of fuel ethanol.[9]

Bt corn growers are the next-largest set of beneficiaries, realizing an estimated $190 million of benefit per year (Figure 3.1) from the greater profitability of growing *Bt* corn. Growers do face somewhat higher prices on *Bt* corn seed, due to the technology fee charged for the *Bt* trait by seed companies, but their profitability is largely driven by their significantly lower expenditures on pesticides. According to another recent analysis, *Bt* corn growers apply fivefold *less* in terms of pesticide active ingredients per acre than growers of conventional corn (National Research Council 2010). Under certain conditions, US corn growers also experience a modest yield gain that helps drive profitability. Together, these factors account for the adoption of genetic technology by growers. As business operations, farms are simply making a rational economic decision.

In fact, the seed industry is able to appropriate only about 15 percent of the total economic benefits generated by the technology, which equals an estimated $128 million per year (Figure 3.1). These are profits based on technology fees. The major providers of *Bt* corn are Monsanto, Pioneer Hi-Bred, and Syngenta. Despite criticism of the monopolistic tendencies of the industry under intellectual property strategies, significantly less economic benefit generated by this technology is captured by the seed industry compared with others down the value chain, including farmers and consumers.

However, there are also costs borne by other stakeholder groups in society, and these are also estimated in Felicia Wu's analysis (2004). The most impacted stakeholder group turns out to be non-adopting corn growers, due primarily to the generally lower market price for corn as a result of the generally higher productivity in corn across the industry because of the *Bt* technology. The increase in the overall supply of corn in the marketplace exerts downward pressure on corn prices, reducing all farmers' earnings when they sell their harvest on the market. Both farmers who grow *Bt* corn and those who do not face these lower prices, but the latter also have to contend with higher costs and/or lower yields. The net impact on non-adopting farmers was estimated to be a loss of over $400 million in 2004 (Figure 3.1).

Finally, the profits of companies manufacturing and selling pesticides are estimated to have been reduced by about $32 million (Figure 3.1). If anything, this is probably an underestimate. Clearly, *Bt* is a substitute technology for chemical pesticides, with competition in the newly broadened market for both biological and chemical pest control technologies.

Social Benefits of New Genetic Traits Due to Their Impact on Productivity of Agricultural Commodities Globally

Global analyses of productivity in several agricultural commodities suggest some potentially important systemic differences between those crops in which advanced genetic traits have been introduced and those in which they have not (Sexton et al. 2009). In particular, global average yield – defined as production (in metric tonnes) divided by area (in hectares) – of corn, soybean, and cotton have been trending upward significantly since the mid-1990s following the first commercial introduction of *Bt* and glyphosate herbicide tolerance traits in these three crops. By comparison, globally the yields of wheat and sorghum – crops in which transgenic traits have not been introduced – have not grown; even though they have experienced some year-on-year volatility, the average has not increased. While this constitutes only circumstantial or indirect evidence, the case can be made that introduction of these genetic traits has been correlated with important productivity gains in corn, soybean, and cotton.

**Other Transgenic Traits with Greater Perceived Social Benefits
That Have Not Been Widely Commercialized**
There are certainly other possible products of agricultural genomics, and
several are expected to intrinsically deliver greater – or at least less controversial – social benefits. These include crops with drought and stress tolerance
traits as well as crops with improved nutritional contents and crops for clean
energy.[10] Genetic improvement of biofuel crops is expected to improve the
environmental impact of biofuel production.[11] Many of these, however, are
so-called orphan traits or orphan crops. Crops that are primarily grown
and consumed by the rural poor or that are grown on a small commercial
scale – and thus represent little or no market value – tend to be "orphaned,"
or neglected by commercial efforts at genetic improvement. There have
long been humanitarian efforts to create applications of genomic and genetic engineering technologies for some orphan crops, particularly for subsistence farmers.[12]

Although such applications do present ample opportunity for generating
social benefits, the net impacts may not be as stark as those discussed above
for biological insect resistance or herbicide-tolerant cropping systems, and
therefore objective economic measures of their impacts may not be as great.
Still, broad public perceptions of the possible social benefits of such products are often more important than reality, particularly when it comes to
shaping policies governing the innovation and commercialization of products from this technology, including intellectual property rights and market
regulations, to which we now turn our attention.

Intellectual Property and Regulatory Policies
Intellectual property and regulatory policies have significant influence over
the rate and direction that innovation takes within the economy. Historically,
these areas of policy have arisen quite separately, with fundamentally different goals and based in fundamentally different areas of law. They interact
in very important ways, however, and we shall see how they can complement and strengthen one another while at the same time they can partially
substitute for one another. Before discussing their interactions, let us consider some of the broad characteristics of each area of policy.

Debates over Intellectual Property Policies in Crop Genetics
To give some shape and contour to the extensive debate over intellectual
property as applied in crop genetics, it may be helpful to recognize three

fundamentally different kinds of arguments for and against the application of IP to genetics and living materials.

First, there are moral arguments, largely based on the concept of rights. Such arguments in support of IP focus on inventors' natural rights to ownership of their own creations, and the right to enjoy the fruits of their own labours, a discussion very much grounded in a Lockean notion of natural rights. We often hear inventors advocating the strengthening of IP rights with arguments along these lines. On the other hand, opponents often raise moral or rights-based arguments against extending IP rights to the subject matter of genetics and living materials, fundamentally arguing that it is immoral to grant broad exclusive ownership over the features of living things, or at least that it introduces potential conflict between different sets of rights. The inventor's intellectual property rights may come into conflict with certain declared basic human rights, such as the right to food security and basic nutrition or the right to basic health care; this is often seen, for example, in debates over humanitarian access to proprietary drugs (see Plomer 2013).

Second, there are pragmatic, utilitarian, or instrumentalist arguments. Much of the debate and analysis that occur within the discipline of economics and other social sciences tend to fall into this category. One example of a utilitarian argument in support of patent policy is the classic trade-off between the granting of a temporary monopoly in exchange for public disclosure of the invention. A utilitarian critique of patent policy is the "anti-commons hypothesis": that the proliferation of intellectual property claims, insofar as it increases transaction costs and strategic behaviour, may reduce in aggregate the incentive to innovate and thus erode the practical aims of the intellectual property system.[13]

Third, there are realist or political economy arguments with regard to intellectual property policy, primarily focusing on the relative incentives, rent-seeking opportunities, and strengths of different interest groups.[14] Such arguments acknowledge that those who innovate have significantly greater incentives to support the protection of innovations. Thus, innovators are more likely to comprise interest groups that advocate stronger IP policies. Such arguments can often be seen at the heart of North/South debates and negotiations over international agreements on intellectual property standards and practices.

Debates over Regulatory Systems

We turn now to those regulatory systems whereby governments require innovators to register and/or receive approval of new products before they may be sold on the market, on the grounds of protecting consumer safety and environmental quality.

One of the basic goals of regulating the introduction of new products is, of course, to standardize, or at least provide a baseline of predictability about, the safety and quality characteristics of new products. There is a whole line of argumentation, however, that government regulation is not the ultimate guarantor of consumer protection, and in some cases may provide little improvement over the discipline of the marketplace. Proponents of this view contend that there are powerful market incentives for companies not to sell products, such as pharmaceuticals, that may harm their customers.[15] This can certainly be seen in the commitment of companies to protecting their brands' reputation for quality and safety, often exceeding regulatory standards that may be in place. While this market assurance likely serves customers quite well where reputation is important and information is easy to share, it does not work in all cases. One example, of course, is tobacco. If a product causes harm less visibly, more indirectly, or simply more slowly, then a *laissez-faire* market is less able to self-regulate and some sort of external control is needed to ensure product standards. Most products are somewhere in the middle, where the purpose of regulation is more the standardization and predictability of product characteristics to marginally improve market efficiencies or overcome market failures.

Another important characteristic of regulatory systems is the basic alignment of incentives, with the innovator bearing the costs of proving the safety or efficacy of their new product in order to obtain marketing approvals, as opposed to the costs being borne by the public. Such a structure is built into most of our environmental and biosafety regulations, largely for budgetary reasons. The cost of ensuring safety or efficacy rests with the innovator rather than with the government. It is essential to note that this alignment effectively raises the fixed costs borne by innovators who want to introduce new products. In markets that are relatively easy to enter with additional versions of a regulated product, this creates a free-rider dilemma and means that initial innovators need to have exclusivity over their products in order to simply earn back the fixed investment that they have to pay but later entrants do not. The classic example of this is the market for pharmaceuticals, where generic versions do not bear a proportionate share of the cost of proving safety and efficacy.

An additional characteristic of regulatory systems is, in effect, the shifting of some liability from the innovator to the regulator. In the United States, for example, if a product is in compliance with regulatory requirements and a consumer safety incident occurs, public accountability (at least), if not legal liability, may be shared between the regulator and the innovator: not everything rests on the shoulders of the innovator.[16]

Another very important characteristic is the scope of approval granted by regulators. This varies across different technologies and markets, but as a general rule specific products or technologies are approved on a case-by-case basis. Data submitted to prove safety and efficacy are applicable only to a defined product or range of products. Pragmatically, this is because the same data may simply not be capable of proving scientifically the safety or efficacy of other products or applications, or such use of submitted data may not be permitted for confidentiality reasons. In some cases, such as with a small-molecule pharmaceutical, a second drug product may be able to come in under an existing approval. If the molecule is identical, and if science predicts that it will behave in exactly the same way, regulatory approval may be extended to the second product (see *Food and Drug Regulations* and *Applications for FDA Approval to Market a New Drug – Abbreviated Applications*). This has yet to occur reliably for biopharmaceuticals and biological products, however, largely because of the complexity of the products and their respective mechanisms of action. Although follow-on products may be designated as being "similar," there is not full portability. As a result, regulatory approvals of biological products in most jurisdictions are restricted to a specific product.

A final characteristic of regulatory approval systems that may prove to be quite significant in crop genetics is the time horizon of product approvals. When approvals are time-limited, there are necessarily requirements for re-application or renewal if the product is to remain on the market. The purpose of including a time limit is to enable regulators to review products regularly, with a built-in mechanism for removing them from the market if problems are found. However, such time limitations raise serious questions about the costs of keeping products on the market even if they have been proven effective.

The International Reach of National Policies

It is important to reflect on the fact that, even though both IP and regulatory systems are established and operate within specific jurisdictional contexts (usually either multinational, such as the European Patent Office, or national, such as the Canadian Intellectual Property Office), their impacts are often broader due to the integration of markets and trade in goods that embody protected or regulated technologies. When it comes to introducing innovations in widely traded commodities such as agricultural products, it is essential to consider the IP and regulatory systems at work within all of the multiple jurisdictions spanned by the effective marketplace. This is, of course, what has driven many of the agreements seeking international harmonization of IP rules and regulatory standards. In agricultural genomics in particular, significant differences remain in the strength of IP protections and the regulatory systems in place in different jurisdictions.

In some cases, these disparate systems reflect legitimate differences in moral or practical concerns within the different jurisdictions, but forces are also brought to bear on both the IP and regulatory systems by the different industries and interest groups that have varying degrees of political influence within those jurisdictions.[17] Innovators thus have to operate in a range of different theatres simultaneously. Each new innovative product will require regulatory clearance in each relevant domestic and export market. IP protections will also vary across the relevant jurisdictions. As a result, innovations using agricultural genomics or biotechnology require an international IP management and regulatory stewardship strategy; innovators can thereby effectively achieve protections globally due to commodity trade flows.

The IP–Regulatory Complex

The typical policy position we tend to see innovators adopt is that strong IP protection is required to ensure sufficient returns to cover the high fixed costs of research and development (R&D) and regulatory compliance. Such language is common in *amicus curiae* briefs in IP cases, such as *Association for Molecular Pathology v. Myriad Genetics, Inc.*, and can be found in the policy positions of industry groups such as the Biotechnology Industry Organization (2013a, 2013b) or the Pharmaceutical Research and Manufacturers of America (2013).[18] This argument follows the utilitarian or instrumentalist line of reasoning about intellectual property. Effectively, it implies a social contract: that a government should provide for strong IP protection if that government is going to impose costly regulatory requirements for obtaining market approvals.

As a result, regulatory approvals and formal IP protections (including patents, plant variety protections [PVPs], trademarks, and so on), together with less formal IP protections (such as the disincentive created by high data costs, exclusivity of data submitted to the regulator, as well as secrecy, non-disclosure, and noncompetition contracts, and so on), provide mutually reinforcing or complementary mechanisms of market exclusion. When both

Figure 3.2
The IP–Regulatory Complex: Regulatory approvals, together with formal and informal IP, provide mutually reinforcing or complementary mechanisms of market exclusivity

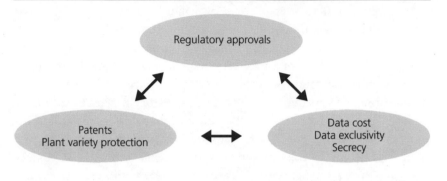

are held by an innovator, IP and regulatory approvals constitute a suite or portfolio of protections that can be developed to maintain control over an innovation. These interact to reinforce market exclusivity (see Figure 3.2).

The market exclusivity maintained by the regulatory aspect of the IP–Regulatory Complex is exemplified in cases where an innovator is able to maintain market exclusivity despite the absence of patent protection. One example is that of *Bt* cotton during the early years of its commercial release in India, particularly between 2002 and 2005. Despite an extensive search of the Indian patent literature, we have not been able to find that the innovators of this technology – Monsanto and its commercial partner in India, Mahyco – held any patent granted by the Indian patent office for the *Bt* cotton technology.[19] From 2002 until 2005, however, Monsanto and Mahyco did occupy the enviable position of being the only company to have received approvals from Indian biosafety regulators for the commercial release of *Bt* cotton varieties (Kathage and Qaim 2011). Since each new *Bt* cotton variety needed to obtain biosafety clearance even if it incorporated the (unpatented) Monsanto trait, the cost to potential competitors of generating a new data dossier to submit to the biosafety regulator served as a practical economic barrier to the rapid market entry of potentially competing varieties. Instead, it proved in most cases to be more cost-effective for a potential competitor to license Monsanto's regulatory data package for use in support of its own submission to the Indian biosafety regulators for approval of its *Bt* variety, as long as Monsanto's licence fee was less than the cost of generating all of the necessary data to support a completely separate and independent application for biosafety approvals. This effectively left control over the development of the technology in the hands of Monsanto and Mahyco, thanks to India's biosafety regulator. Only in 2006 did two competing *Bt* technologies – independently developed by Indian and Chinese academic scientists, respectively – legally enter the market.[20]

Another example of IP–like control being bestowed by the regulatory system can be seen in the case of glyphosate-tolerant (Roundup Ready) soybeans. The last of the portfolio of patents that protected this technology expired in Canada in 2014 and is set to expire in the United States in 2015. Once off patent, the technology would, in theory, enter the public domain and be used to create generic products, thus finally fulfilling society's side of the traditional deal struck under patent law between society and inventor. Under this deal, society grants the inventor a temporary monopoly in exchange for the inventor's disclosure of the technology to society, so that once the patent expires the practice of the technology enters the public domain, instead of being held in secrecy, or at least relative obscurity, indefinitely.

The catch here is that in several export markets, including Europe and China, regulatory approvals of genetically engineered crop varieties are temporary by design. If renewal applications are not submitted, the regulatory approval will expire and it will become illegal, under current rules, to sell the product with the off-patent technology within that jurisdiction. Moreover, when its patents expire, the original innovator (in this case Monsanto) loses any incentive to incur the additional cost of submitting applications for regulatory renewals with the European or Chinese biosafety authorities. To make matters worse, the off-patent technology will be competing with Monsanto's new product, Roundup Ready 2, which means that Monsanto would have every incentive to allow the European and Chinese regulatory approvals of the original Roundup Ready technology to lapse.

Soybean growers, even in the United States, Canada, and other countries where the original technology was approved, will not grow a crop that cannot be exported, because of the costs of differentiating separate streams within the massive global flows of this commodity and the risks that expiration of regulatory approvals in a major export market would cause major trade disruptions. A competitor could decide to incur the costs of repeating the trials and replicating Monsanto's biosafety data in order to support the reapplications for regulatory renewals in China and Europe, but would all of the potential vendors of different versions of "generic" Roundup Ready soybeans be so willing? Again, it may just be more economical for each to license the existing data dossier directly from Monsanto. Monsanto maintains exclusive control over the technology, this time post-patent, thanks to the regulatory regime.

In fact, Monsanto has pledged publicly to maintain worldwide regulatory approvals of the original Roundup Ready soybean through 2021 (Monsanto 2014). However, this case raises important questions regarding the legal pathway to achieve approval for an agbiogeneric, particularly the effective exclusivity created by the current regulatory system. Effectively, regulatory requirements leave control over an off-patent technology in Monsanto's hands.

Effects on Incentives to Develop New Products

We have argued that regulatory and IP systems are complementary and, as currently constructed, appear to reinforce or extend the control of a few innovators. Now we ask: What are the effects of this IP–Regulatory Complex on innovation and commercialization of socially beneficial agricultural products from agricultural genomics, or products where, in fact, there may be less private profit incentive to take the product to market in the first place?

Effects on the Development of Drought-Tolerant Crops

We have recently completed an analysis of genetically engineered drought-tolerant crop varieties in the R&D pipeline globally, surveying research articles, patent families, field trial data, and regulatory submissions in every major jurisdiction except China (Graff, Hochman, and Zilberman 2013). Around the world, governments have invested massive resources to understand the genetics of stress and drought adaptation by plants, which is evident in the exponentially growing number of publications we observe in the field. Both public sector and private sector entities have been filing patents on drought-tolerance genetic technologies in the major patent jurisdictions around the world, demonstrating that some applied technologies are indeed emerging from the science and companies are beginning to see this as a potentially promising market. To date, however, there have been very few regulatory filings anywhere in the world: just a handful in the last three years, all covering a single genetically engineered drought-tolerant corn variety.

Without a benchmark for comparison, these trends in the R&D pipeline – with scientific research coming first, followed by a rise in patents, then field trials, and finally a handful of regulatory submissions – may simply reflect natural dynamics, with a lag between the science and its commercialization. Perhaps the commercialization of crops endowed with drought-tolerance traits will take off in the next few years, vindicating the aforementioned public investment in early-stage basic plant genetics and genomics and the functionality of the IP–Regulatory Complex.

There is more to this story than just genetically engineered drought tolerance, however. Three drought-tolerant corn varieties are being introduced to markets in the United States. Two are conventionally bred (introduced by DuPont Pioneer and by Syngenta), while only one is transgenic (introduced by Monsanto). In practice, in the United States, conventionally bred varieties are not subject to the same regulatory requirements as genetically engineered transgenic varieties (Marden 2003), but they still require significant investment and expertise in genetics.[21]

Globally, the scientific research in drought tolerance is fairly evenly divided, with Europe, the United States, other Organisation for Economic Co-operation and Development (OECD) countries, and developing countries each generating about 25 percent of the basic scientific publications in drought-tolerance genetics. This even pattern, however, is not reflected in the global patent filings. A review of the patents in the field of crop drought-tolerance genetics reveals that the commercial leaders in the field are DuPont Pioneer (US), Monsanto (US), BASF (Germany), Syngenta (UK/Switzerland), and Bayer (Germany).[22] Public sector entities – perhaps not surprising given the magnitude of public science in the field – account for close to a third of the patented inventions in this area.

In conducting field trials of drought tolerance, corporations are concentrating on particular crops, giving an indication of what they are likely to bring to market. Corporations are particularly focused on field trials of drought-tolerance traits in grains and in oilseeds, primarily soybean. But that is the extent of it: corporations are not developing drought tolerance in any other crops for the market. By contrast, public research institutions and universities are doing field trials of all kinds of crops, including drought-tolerant grains and oilseeds, as well as drought-tolerant horticultural crops and forage and forestry species. While these may be complementary emphases of R&D, the pattern does raise questions regarding whether any drought-tolerant crops other than those major grains and oilseeds on which corporations are focused will ever be brought to market.

Effects on the Development of Products with Improved
Nutrition and Other Qualities

The global R&D pipeline of transgenic nutritional and product quality traits like vitamin A rice, high-lysine corn, and longer shelf life vegetables – another area of innovation generally presumed to promise a wealth of socially beneficial product technologies – likewise exhibits an interesting and potentially disturbing trend attributed to the interacting effects of IP and regulation (Graff, Zilberman, and Bennett 2009, 2010). Aggregating annual data on scientific publications, field trials, and regulatory filings has shown that the growth in the overall number of innovations in the R&D pipeline levelled off abruptly – as if it had hit some sort of glass ceiling – in 1998, the same year that the European regulatory approvals process effectively went into moratorium (Graff, Zilberman, and Bennett 2009). Statistical analysis of that change in the innovation growth curve shows a highly significant shift in parameters around 1998, signifying a fundamental break in trend. A deeper look at the relative dynamics of scientific articles, field trials, and regulatory filings shows a decrease after 1998 in the rate and likelihood with which innovations were both being introduced and moving down the R&D pipeline (Graff, Zilberman, and Bennett 2009). Also, in the later years, after the additional regulatory hurdles were introduced, a greater share of field trials was being undertaken by public sector researchers; or rather, conversely, a reduced share was being conducted by industry innovators. In other words, commercial expectations had dropped in response to changes in the regulatory environment.

In addition, there was a change in the composition of companies engaged in this R&D. Those actively working on nutritional and product quality innovations during the earlier period, from 1992 to 1998, included a fairly diverse set of seventeen seed, agricultural chemical, and food companies. In the latter period, after 1998, the set of companies actively innovating had narrowed down to the same five major agricultural chemical companies noted earlier (Monsanto, DuPont Pioneer, Syngenta, BASF, and Bayer) plus two outliers: Scotts, which was primarily working on higher-quality lawn turf grasses in close partnership with Monsanto, and Vector Tobacco, which was solely focused on developing nicotine-free tobacco products.[23]

In sum, we hypothesize that the pipeline dried up in the later years largely because the relative cost-benefit calculation for smaller-market products fundamentally shifted as patent control consolidated in the hands of five major companies and as regulatory approvals became much more costly to attain. We cannot attribute the shift fully to IP constraints, nor can we pin it just on regulatory constraints. Rather, we believe that what we are seeing is due to the interaction of the two.

How the IP–Regulatory Complex Shapes Incentives to Develop New Socially Beneficial Products

Based on these studies and other observation of this industry, we can discern several impacts that the combination of IP and regulatory systems appears to be having on innovation in crop genetics and genomics.

Skews Innovation toward Applications for Large Markets

The first and most obvious impact is that the high costs of regulatory approvals combined with the attainment and maintenance of costly IP portfolios skew innovation toward large markets. This is similar to the dynamic in the pharmaceutical industry that favours development of blockbuster drugs while leaving diseases that affect smaller populations "orphaned." Not only are innovators drawn by large incentives but they also face very large costs. All of these mean that only the tip of the iceberg of technically possible innovations – just those at the head of the distribution of expected payoffs – actually makes it to commercialization in the marketplace. At the same time, this pulls innovators away from smaller markets, including specialty products for rich-world consumers as well as crops produced and consumed by the subsistence poor in places like India and sub-Saharan Africa.

Importantly, for those in the land grant universities in the United States and similar agricultural universities around the world, the large minimum investment required to pursue development of these technologies effectively eliminates the use of transgenic technology to address regionally specific agronomic problems. Even if there might be an elegant transgenic fix for a particular pest that afflicts a state or provincial growing region, unless circumstances are exceptional (for example, if the region is the only reliable source globally of a particular product), the public sector research into and development of a transgenic technology will most likely not proceed. Regional growers will not be willing to make the investment in IP and regulatory costs or bear the regulatory and associated market-acceptance risks in foreign markets.

Reinforces Oligopolistic Control of Technology Platforms

We also argue that this situation reinforces oligopolistic control of technology platforms. Developing a technology platform is part of a much more complex innovation capacity-building process that takes place at large companies. They accumulate tacit human capital, know-how, and other complementary R&D assets largely because the incentives align for them to be building capacity at such a scale. The incentives are simply not as powerful for smaller biotech entrepreneurs or even for universities to be building up the integrated product development, IP, and regulatory compliance capacity to move products to market, because these entrepreneurs and universities will simply not be able to service the largest markets. This also drives a product mix that serves the shareholder interests of a few large firms, focusing on product categories where expected returns are commensurate with the stock market rate of return.

Creates Incentives for Incumbent Firms Not to Innovate, or to Delay Introduction of Innovations

The degree of control held by those who have a product on the market can create incentives for incumbents not to innovate, or at least to delay improvement or replacement of existing (protected and approved) products. This is in the spirit of the "innovator's dilemma" argument (Christensen 1997): why would a company that has a great product without close competition in the marketplace seek to displace it with something else? The optimal timing of investment to sink capital into creating something new and attaining regulatory clearance falls closer to the end of the existing product's patent life. For example, Monsanto waited to introduce a fundamentally new and improved genetic construct for glyphosate resistance, its Roundup Ready 2 Yield, until just a few years before the patent on its original Roundup Ready technology expired.

Extends Control of the Technology beyond the Life of the Patent

The combination of regulatory requirements and IP protection may effectively extend control beyond the life of the patent, and effectively thwart the entry of the technology into the public domain. As discussed in the example of Roundup Ready soybeans, the biosafety data required by regulators for renewal of approvals may deter subsequent entrants into the market. Consequently, the first entrant into the market may provide the only legally available product, and maintain exclusive control over the regulatory data necessary to seek approval, even after the patent has expired.

Reduces Diversity of Sources of Innovation by Consolidating the Structure of Industry and Discouraging Entrepreneurship

There is a tremendous amount of great research going on in the plant science laboratories of such universities as the University of British Columbia, Colorado State, Ohio State, Cornell, and the University of California at Berkeley, Davis, and elsewhere, but commercial developments from those research efforts that rely on transgenic technology are highly unlikely unless one of the aforementioned five big "gatekeeper" companies finds it useful. There is very little in the way of a viable entrepreneurial community in agricultural biotechnology today. Venture capital is not interested in investing in a technology unless there is expressed interest, or even an existing agreement with one of the five companies. Overall, as we have tried to show empirically in the areas of drought tolerance and nutritional genetic traits, this has probably already curtailed the rate of innovation across the entire industry.

Of course, innovation is always a highly uncertain undertaking. Out of hundreds of ideas and prototypes, only a few will succeed. In a young industry, where there are large potential social benefits to be gained across a broad suite of possible products, technologies, and markets – both developing and developed – the lack of uptake by large companies can be, and likely has been, detrimental to social benefit. The world is not seeing the kind of exponential growth in innovation that might have been possible, because of the sheer scope of thwarted technological opportunity.

Shifts R&D toward Use of Nontransgenic Methods Such as Mutagenesis and Molecular Breeding

A final note is that IP and regulatory barriers may be inducing a shift toward the use of nontransgenic approaches to improving crop genetics. All else being equal and if it is technically feasible, developers are more likely to focus their efforts on techniques such as mutagenesis and molecular breeding, because costs along the path to market in the United States, Europe, and many other jurisdictions are so much lower than those for transgenic technologies.

Proposed Solutions

Questions about how IP and regulatory systems can be restructured to maximize innovation for socially beneficial applications have been explored extensively, particularly in the fields of pharmaceuticals and agriculture.[24] Quite a number of interesting proposals have been advanced, suggesting a variety of mechanisms. Several of these could, in our view, improve the structure of the interdependent intellectual property and regulatory approvals systems. Others, we expect, are less likely to make much of a difference.

More often than not, this is because the proposed fix is too narrowly focused on just IP or just regulatory approvals, looking at only part of the overall problem while failing to consider interactions between the two. In short, it is crucial that reform proposals for either IP or regulatory regimes acknowledge the nature of the IP–Regulatory Complex.[25]

IP Solutions

A number of proposals for IP law and policy, as well as for what we might call IP "private ordering" or "collective rights" mechanisms, have been made in the literature and public arena. An IP reform proposal that has garnered much public attention has been for a subject matter exemption against patent claims over isolated genomic DNA. This proposal has been advanced in the United States by a pair of Supreme Court decisions, in *Mayo v. Prometheus* (for methods claims) and *AMP v. Myriad Genetics* (for composition of matter claims). This exemption from patentability of genomic DNA has been promoted as a way to stimulate innovation by enabling broad use of naturally occurring genomic sequences by innovators, while still awarding IP exclusivity over non-natural or modified genetic constructs. We expect this to remove some perceived barriers to early-stage innovation and commercial use in a few smaller-market applications, but it is not clear to what extent this reform also undermines incentives for innovators trying to make it through regulatory approvals.

A second class of IP proposals have called for the tightening of patentability criteria, including utility, novelty, and nonobviousness standards, as well as requirements for providing written description or reduction to practice – or at least the tightening of their interpretations and applications when it comes to genetic inventions. Perhaps the most visible example of such a reform was the introduction of the *Utility Examination Guidelines* by the US Patent and Trademark Office in 2001, significantly raising standards for claiming genetic sequences (see *Interim Utility Examination Guidelines* and *Utility Examination Guidelines*). When well crafted, such reforms generally result in better-quality patents. Although they potentially reduce the total volume of patents issued, those patents that are not issued because of these reforms probably never should have been, while those that are issued are more likely to represent truly inventive contributions. The resulting IP protection is generally stronger, giving innovators more protection and thus greater incentives to invest in obtaining regulatory approvals.

A third class of IP proposals advocate research-use exemptions – as well as other public-use, government-use, and even humanitarian-use exemptions

– and compulsory licences. These policy changes are not concerned with what should or should not be eligible for patenting. Rather, they accept what is patented as given, and then seek to selectively allow certain activities that would otherwise infringe granted patents – at least for uses that pass some test of being "for the public good." The mechanism works by simply curtailing the ability of the patent holder to exclude those designated uses. Further, in the case of compulsory licensing, the patent holder is given some financial compensation.

In general, this category of proposals seeks a compromise between the interests of the inventor, preserving the underlying property right award, and the interests of those who wish to use the invention for designated uses that are, presumably, in the public interest. While this compromise sounds laudable, the feasibility of such policies hinges on the workability and enforceability of the test or definition of the designated public use. In practice, the drawing of such lines is never as clear as the advocates of such reforms envision. Moreover, the owners of patents that are potentially subject to such government use or compulsory licence actions have strong private incentives to appeal any decisions made under these clauses. In the United States, government "march-in" rights over patents resulting from publicly funded research, as provided for under the *Bayh-Dole Act*, have virtually never been exercised (except in one or two dubious cases, where the action appeared to have been motivated more out of political spite that out of public interest; see O'Connor, Graff, and Winickoff 2010). In Canada, compulsory licence provisions have likewise languished largely unused.[26]

One additional category of IP proposals includes what some legal scholars call "private ordering" arrangements, essentially privately organized collective action solutions (O'Connor, Graff, and Winickoff 2010). Like the previous category, these proposals do not challenge patentability requirements and accept granted patents as granted. They do, however, seek to improve the efficiency of the market for accessing rights to use technologies protected by patents once they are granted – by reducing information asymmetries and search costs, reducing transaction costs, or reducing opportunities for strategic bargaining. There is a range of such proposals, including calls for provision of better information on patents and patented technologies, such as clear and systematic data on which nucleotide sequences are referenced or claimed in patents; calls for consolidated patent licence listing services or clearinghouses, facilitating low-cost market discovery and licensing transactions; and calls for higher-order patent-pooling or platform licensing arrangements, leading to wider access to enabling technology platforms.[27]

Notably, all four types of proposed IP solutions involve some sort of weakening of the IP rights, or at least of the strategic advantage enjoyed by the holders of the IP rights. Those rights holders, who have made private investments in the uncertain and risky world of genetics and genomics innovation, are not likely to be enthusiastic about such proposals to reduce their probability of earning returns on their investment. From a societal point of view, however, some reduction in the incentives enjoyed by paten- tees may be warranted if it means that improved access to their IP–protected technologies leads to a commensurate increase in socially beneficial appli- cations of the technology. This prompts a critical question, however: for those applications of genetics and genomics that face significant regulatory requirements, who is going to fund the risky undertaking of the develop- ment and regulatory approvals process? For, all else being equal, private in- vestment is likely to respond to a weakening of incentives to invest as expected. To what extent are taxpayers and public agencies, or foundations and donors, ready to step into the breach? Are they in a tenable position to sink significant amounts of risk capital into the development of products that may fail? In fact, even if public or social enterprise financing may ad- vance flagship projects, we are unlikely to find a whole armada of publicly funded development projects in their wake. The efficacy of IP reform in freeing up the innovation pipeline thus depends, to a significant extent, on stimulating investment in follow-on innovation, but in a weaker IP environ- ment, this necessarily hinges on reform in regulatory requirements.

Regulatory Solutions
A number of proposals have also been advanced for reforms of regulatory policy. The first major proposal is to rationalize the regulation of transgenics relative to other methods of induced genetic variation in crops, and, in so doing, to shift the object of regulation from a genetic transformation "event" to a variety's phenotype (Bradford et al. 2005; Potrykus 2010). This proposal argues that methods are not relevant to assessing the objective level of risk posed to environmental quality or human health. Rather, what matters from a risk management perspective is the nature of the crop traits as physically expressed in the environment or in the food product. In other words, what matters is not how genetic variation is achieved but rather what that varia- tion produces. Thus, regulatory requirements should not focus on the method of creating the genotype but rather on the resulting phenotype.

This stands up to logical scrutiny. After all, plant varieties with genetic traits comparable to those achieved by genetic engineering are common in nature and have long been used in conventional breeding programs. Insect resistance is a naturally occurring phenomenon in many plants, with genes expressing phytotoxins within the plant tissues that are naturally evolved defences against pest predation. Many of these natural biopesticidal traits have high target specificity, similar to the *Bt* trait, and have been selected by plant breeders over the decades and bred into commercial crop varieties to increase resistance to pests. Indeed, without them we would have much lower yields in many crop species. Similarly, tolerance to herbicides has been induced in individual plants by mutation and then bred into commercial crop varieties. Some of these are currently marketed as competitors to the transgenic herbicide-tolerant cropping systems, but, interestingly, because they are not considered "transgenic," they are not subject to regulation directed at transgenic technology.[28] For many such traits introduced to crops using transgenic techniques, there may not be any greater risk or need for regulatory oversight than there would be for any other newly bred crops with novel traits. The proposal therefore is to treat all novel genetic technologies equitably under regulatory requirements, essentially levelling the playing field and rationalizing the cost of compliance so that it is commensurate with the objective risk posed by new products.

This brings us to a related proposal for the introduction of a spectrum of tiered risk categories, with increasingly stringent requirements for increasingly risky technologies, a regulatory approach common in environmental toxicology (see Bradford et al. 2005; Romeis et al. 2008). Such an approach has to a certain extent been adopted by the Canadian Food Inspection Agency (2004) with regard to plants with novel traits. The greatest advantage of the tiered system is that it aligns the cost of compliance more closely with the actual risk posed. In particular, it allows for relatively low-risk innovations to obtain approvals at relatively low cost. This in turn reduces the need for strong IP rights, and both effects simultaneously reduce barriers to the entry of additional innovations that will compete in the same market.

A final policy approach has been to simply subsidize or expedite regulatory approvals for select technologies, particularly those for small or underserved markets. This has not been common for transgenic crop varieties or for novel plant traits, but one example can be found in the US Department of Agriculture's IR-4 Minor Crop Pest Management Program, which provides subsidies to support the regulatory approval of pesticides for small-market specialty crops (National Institute of Food and Agriculture 2014).

Combined IP and Regulatory Solutions

Given the interdependence between IP and regulatory systems, we argue that the greatest potential for affecting the incentives to develop socially beneficial products from agricultural genomics lies in the integration of IP and regulatory solutions.

Generally, from a policy maker's perspective, this can be achieved by combining any of the separate IP and regulatory solutions discussed above. For example, the introduction of stronger patent eligibility requirements, which would reduce the number of superfluous granted patents in the field and thus the general risk of infringement by follow-on innovators, can be combined with a rationalized regulatory registration system, thereby reducing compliance costs. These two changes made in combination would do significantly more than either one alone in terms of improving incentives for innovation.

Orphan drug policies comprise one area that blends both IP and regulatory aspects.[29] The main thrust of orphan drug policies is to combine stronger IP terms – for example, by extending patent life – with expedited or subsidized regulatory review processes. Other incentives may be added as well, such as R&D tax credits. Although several countries have adopted orphan drug policies, we are not aware of any instance where the model has actually been applied to the genetic improvement of "orphan" crops.

In addition to integrated IP and regulatory policy reforms, there are collective action initiatives that seek combined approaches to solving current IP and regulatory challenges in the interest of advancing socially beneficial applications of agricultural genetics. The Public Intellectual Property Resource for Agriculture (PIPRA) was established in 2003 at the University of California at Davis to provide public sector researchers and their institutions with capacity support for navigating and managing the IP and regulatory aspects of technology transfer and commercialization of crop genetic technologies (see Atkinson et al. 2003; Bennett et al. 2008). In India, the Platform for Translational Research on Transgenic Crops (PTTC) was established on the campus of the International Crops Research Institute for the Semi-Arid Tropics (ICRISAT) in Hyderabad, with a mission to help manage late-stage development as well as IP and regulatory needs for commercialization of crops important in Indian agriculture (Sharma 2010).

In our view, these represent a potentially helpful approach, insofar that they seek to engage public sector agricultural research institutions more fully as innovators, building on their existing research capacity and their long history of creating socially beneficial applications for agriculture, particularly filling those gaps where incentives are insufficient to entice commercial enterprises. However, simply providing additional IP and regulatory capacity on top of the underlying noncommercial mission and capacities of existing public sector research institutions is insufficient to shift the incentives such that socially beneficial products will enter the market. Both PIPRA and PTTC acknowledge this and seek to develop partnerships with seed companies or other commercial enterprises in a position to make the final push toward commercial release of new products.

Among their potential partners is another type of collective action initiative that has become quite prevalent – both in developing drugs and vaccines for neglected diseases as well as in developing crops with advanced genetics for subsistence agriculture – known as the public/private product development partnership. These operate essentially like firms, except that as nonprofit organizations they are able to uniquely combine R&D funding from governments, foundations, and philanthropists *and* from commercial entities. They can also sell products or license technologies and plow the resulting revenues back into their R&D efforts. These organizations take full stewardship of the product development process, managing IP access as well as clinical or field trials through the entire regulatory approval process. Public/private product development partnerships often inherit or transfer promising technological leads from a commercial or a university partner and then seek to complete the development of a product, be it a drug or an improved crop variety. One prominent example of such a partnership in agriculture is the Water Efficient Maize for Africa (WEMA) initiative, which brings together Monsanto, the African Agricultural Technology Foundation, the International Maize and Wheat Improvement Center (Centro Internacional de Mejoramiento de Maíz y Trigo, or CIMMYT), and national agricultural research systems in several African countries. The WEMA partnership consists of a product development team, an intellectual property team, and a regulatory affairs team, and began releasing hybrid maize varieties under licences for commercial production in Africa beginning in October 2013 (Kitabu 2014).

Improving Incentives for the Innovation of Socially Beneficial Products
In sum, IP and regulatory systems, although separate in terms of policy goals and administration, interact significantly to govern the incentives to innovate. The greatest degree of market exclusivity and thus incentives

come from regulatory approvals, largely due to the high fixed costs of application and compliance. Although it is often argued that IP protections are necessary in order for innovators to be willing to undertake the risky investments to pursue the costly regulatory approvals process, it may be that the attainment of regulatory approval is its own reward, since, particularly in the area of biologicals and crop varieties, the free-rider problem from follow-on products is all but nonexistent; instead, regulatory approvals can end up reinforcing the market exclusivity attained with IP rights. Even in the absence of IP protections, regulatory approvals can effectively keep generic competitors out of the market.

This synergy between IP protection and the protection attained under regulatory approvals, we argue, has significantly reduced the rates of innovation across a range of applications of agricultural genomics that would be considered socially beneficial. If anything, the current alignment of policies reinforces the position of large incumbent corporate innovators and locks out small innovators.

We therefore argue that solutions require coordination between IP and regulatory reforms. Most of the IP reforms that have been proposed to date seek a relative reduction in the strength of IP rights, but this is likely to have little effect in the shadow of an unreformed regulatory system; if anything, it will harm the smaller innovator while leaving the large players relatively untouched. An ideal combination for incentivizing innovation would consist of both a rationalization of regulatory requirements – making requirements commensurate with objective risk and thereby reducing costs on most innovations – and improvements in the quality of IP rights and the provision of private ordering infrastructure to increase the transparency and efficiency of IP licensing markets.

Notes

1 Life-cycle analysis is an engineering methodology used to calculate the total cumulative environmental impact that results from the production and consumption of a given product, beginning with the production and extraction of input materials, manufacturing, transport, operation, and disposal. LCAs account for energy and water requirements, pollutants, greenhouse gas emissions, and more. See Hendrickson et al. 2006.

2 "Personal health": A recent meta-analysis of clinical studies conducted by researchers at Stanford Medical School found that organics are not statistically significantly any more nutritious or healthful than conventional foods (see Smith-Spangler et al. 2012). "Social or environmental benefits": A recent meta-analysis of relative yield performance of organic and conventional farming systems published in *Nature*

found that organic yields are typically lower than conventional yields by 5 to 34 percent, signifying that organic products require more land and other resources (see Seufert, Ramankutty, and Foley 2012).

3 For a detailed review of the evidence, see Chapter 2, "Environmental Impacts of Genetically Engineered Crops at the Farm Level," in National Research Council 2010.

4 *Bt* is shorthand for *Bacillus thuringiensis,* a microbe that is the natural source of the pesticidal protein. The gene from *Bt* was identified and transferred to crop plants to enable them to also produce the pesticidal protein. Applications of *Bt* microbial solutions and extractions are commonly used as a biopesticide in organic agriculture.

5 For exposition of these arguments, see Kloppenburg 1988; Magdoff, Foster, and Buttel 2000; and Altieri 2004. For a recent critique, see Lynas 2013.

6 See, e.g., the "Non-GMO Project," an initiative that provides a product package label for food manufacturers along with a verification system to ensure that their products do not contain genetically engineered crop ingredients. See Non-GMO Project 2014.

7 Studies have found contradictory results when attempting to assess consumer valuation of genetically engineered foods. See House et al. 2004; Lusk et al. 2004; and Costa-Font, Gil, and Traill 2008.

8 The Colorado survey on attitudes about agriculture and food was first administered in 2001 and updated in 2007 and 2011. See Sullins et al. 2012.

9 This analysis was conducted before the 2007–13 surge in commodity prices. While consumers experienced significantly higher prices for corn as well as these other goods whose prices reflect the cost of corn during the commodity price surge, it stands to reason that the increase in food prices would have been even greater during the run-up in commodity prices without the greater productivity realized by *Bt* technology.

10 For a review of stress tolerance crop traits in the global R&D pipeline, such as drought tolerance, salinity tolerance, and temperature stress tolerance, see Graff, Hochman, and Zilberman 2013. For a review of nutritional and product quality crop traits in the global R&D pipeline, such as protein, vitamin, or omega-3 fatty acid content, see Graff, Zilberman, and Bennett 2009.

11 Sexton and colleagues (2009) explain that worldwide projections by the Food and Agriculture Organization of the United Nations (FAO) and the International Food Policy Research Institute (IFPRI) require that 60 percent of production growth required to meet increasing food, feed, and biofuel demand will need to come from improved yields; however, yield gains in major grain crops have been slowing in recent years. Under these conditions, they argue, genetic improvements are all but essential to achieve the needed productivity growth. For further discussion, see Chapter 1.

12 Such efforts have largely been pursued by public sector research institutions, including public universities, national agricultural research services, and the international agricultural research centres funded by the Consultative Group for International Agricultural Research (CGIAR). For one example of such work, see Varshney et al. 2011.

13 The anticommons hypothesis was framed in reference to property rights more generally by Michael A. Heller (1998) and applied to intellectual property by Heller and

Rebecca Eisenberg (1998). For further discussion of the potential effects of inadvertent disincentives in IP policy, see Chapter 9.

14 There is an extensive "political economy" literature at the interface between economics and political science. George Stigler (1971) was one of the earliest to explore the way regulators respond to economic interests; Gene Grossman and Elhanan Helpman (2001) formalized and extended the analysis to a wide range of policies. The current authors extended these lines of thinking to the political economy of intellectual property in Graff and Zilberman 2007.

15 In a classic study of pharmaceuticals, Sam Peltzman (1974) argues that selective pressures of the market do serve to protect public safety. The argument rests on the premise that, to the extent a product causes observable harm to users or fails to function properly, it will be quickly eliminated from the market by consumer decisions to avoid that product. Further, long-term damage to the seller's reputation outweighs any short-term profits the seller might make from introducing the substandard product. In light of these market pressures and incentives, he goes on to make the case that strict regulatory requirements for product approval can in fact hurt the greater public interest by creating systemic disincentives against the introduction of new, beneficial innovations. In short, the forgone benefits outweigh the realized gains in safety.

16 This is, we might argue, more generally true in the broader sense of political and economic repercussions than in the narrower sense of legal liabilities.

17 For example, we argue elsewhere that in Europe, Japan, and some developing countries, it has been in the rational self-interest of several concentrated economic interest groups, including incumbent agrochemical manufacturers and farm groups, to oppose the introduction of genetically engineered crops, because they stand to lose sales and profitability if forced to compete against the new technology. These groups therefore exert political influence pressure on their governments to implement regulations that block or slow the introduction of genetically engineered crops. See Graff and Zilberman 2007 and Graff, Hochman, and Zilberman 2009.

18 In *Association for Molecular Pathology v. Myriad Genetics, Inc.*, see *amicus curiae* briefs of Biotechnology Industry Organization (25–27), Boston Patent Law Association (19–23) citing *Kewanee Oil Co. v. Bicron Corp.*, 416 U.S. 470, 480 (1974), 19 ("The patent laws promote this progress by offering a right of exclusion for a limited period as an incentive to inventors to risk the often enormous costs in terms of time, research, and development"), and Intellectual Property Owners Association (21–26).

19 This was corroborated by colleagues who do extensive research on biotechnology in India. We were not able to verify with Monsanto or Mahyco directly.

20 For further details on the interaction of IP and regulatory regimes on *Bt* cotton in India, see Chapter 2.

21 In contrast, under Canadian law, plants with novel traits (PNTs) require regulatory approval, regardless of the process used to create the plant. For further information, see Canadian Food Inspection Agency 2007, which provides a good overview.

22 The five companies account for almost 50 percent of the patented inventions in this area.

23 Vector Tobacco's attempt to develop a "socially beneficial" tobacco product using agricultural genomics provides a whole political economy story in its own right.

First, Vector was subjected to a variety of regulatory challenges from major tobacco companies such as Philip Morris, which felt threatened by the new genetically modified tobacco. Then Vector was challenged by GlaxoSmithKline and other pharmaceutical companies, which alleged that Vector's nicotine-free cigarettes were effectively being sold as a smoking cessation device without having gone through the proper Food and Drug Administration (FDA) evaluation and approvals. See Davis 2003.

24 This section draws liberally on John Barton's work, while recognizing that not all of his discussions about the pharmaceutical industry are equally relevant to agriculture. See Barton and Emanuel 2005; Commission on Intellectual Property Rights 2002.

25 We will be the first to acknowledge that some of our own contributions to this discourse suffered from such shortcomings. We have learned. For example, see Graff and Zilberman 2001, or Graff et al. 2003.

26 While compulsory licences are available in instances of patent abuse pursuant to para. 66(1)(a) of Canada's *Patent Act,* the use of such licences decreased in the 1990s, when amendments to the *Patent Act* coincided with Canada's ratification of the *North American Free Trade Agreement* (NAFTA) and the *Agreement on Trade-Related Aspects of Intellectual Property Rights* (TRIPS Agreement) (Reichman and Hasenzahl 2003, 20).

27 "Data on which nucleotide sequences are referenced": see, e.g., the Hinxton Group (2010, 2012) in reference to stem cells in patents, and Jefferson et al. 2013 in reference to nucleotide sequences in patents. "Clearinghouses": see Merges 1996 for a general treatment, and Graff and Zilberman 2001 for a proposal specific to agricultural genetics.

28 A prime example is the Clearfield herbicide resistance system developed and marketed by BASF.

29 Diseases that only affect small populations or populations with low incomes are "orphaned" by drug and vaccine developers precisely because the market is so small that incentives are weak.

References

Legislation
Applications for FDA Approval to Market a New Drug – Abbreviated Applications, 21 C.F.R. 314, Subpart C, §314.92–314.99.
Food and Drug Regulations, C.R.C., c. 870, Part C, Division 8.
Interim Utility Examination Guidelines, 64 Fed. Reg. 71440 (December 21, 1999).
Patent Act, R.S.C. 1985, c. P-4.
Utility Examination Guidelines, 66 Fed. Reg. 1092 (January 5, 2001).

Cases
Association for Molecular Pathology v. Myriad Genetics, Inc., 133 S. Ct. 2107, 186 L. Ed. 2d 124, 106 U.S.P.Q.2d 1972 (2013), *amicus curiae* briefs of Biotechnology Industry Organization, Boston Patent Law Association, and Intellectual Property Owners Association.
Mayo v. Prometheus, 132 S. Ct. 1289 (2012).

Secondary Sources

Altieri, Miguel A. 2004. *Genetic Engineering in Agriculture: The Myths, Environmental Risks, and Alternatives.* Oakland, CA: Food First Books.

Atkinson, R.C., R.N. Beachy, G. Conway, et al. 2003. "Intellectual property rights: Public Sector Collaboration for Agricultural IP Management." *Science* 301 (5630): 174–75. http://dx.doi.org/10.1126/science.1085553.

Barton, John H., and Ezekiel J. Emanuel. 2005. "The Patents-Based Pharmaceutical Development Process: Rationale, Problems, and Potential Reforms." *Journal of the American Medical Association* 294 (16): 2075–82. http://dx.doi.org/10.1001/jama.294.16.2075.

Bennett, A.B., C. Chi-Ham, G. Graff, et al. 2008. "Intellectual Property in Agricultural Biotechnology: Strategies for Open Access." In *Plant Biotechnology and Genetics: Principles, Techniques and Applications,* ed. N. Stewart, 325–42. New York: J. Wiley and Sons. http://dx.doi.org/10.1002/9780470282014.ch14.

Biotechnology Industry Organization. 2013a. "Intellectual Property." Biotechnology Industry Organization, http://www.bio.org/category/intellectual-property.

–. 2013b. "BIO Commends Supreme Court's Support for Continued Biotech Innovation." Biotechnology Industry Organization, http://www.bio.org/media/press-release/bio-commends-supreme-courts-support-continued-biotech-innovation.

Bradford, Kent, Allen Van Deynze, Neal Gutterson, et al. 2005. "Regulating Transgenic Crops Sensibly: Lessons from Plant Breeding, Biotechnology and Genomics." *Nature Biotechnology* 23 (4): 439–44. http://dx.doi.org/10.1038/nbt1084.

Buttel, Frederick H., Fred Magdoff, and John Bellamy Foster, eds. 2000. *Hungry for Profit: The Agribusiness Threat to Farmers, Food, and the Environment.* New York: Monthly Review Press.

Canadian Food Inspection Agency. 2004. *Directive 94–08: Assessment Criteria for Determining Environmental Safety of Plants with Novel Traits.* Ottawa: CFIA.

–. 2007. *Regulation of Agricultural Biotechnology in Canada: A Post-Secondary Educator's Resource.* Ottawa: CFIA. http://publications.gc.ca/collections/collection_2007/cfia-acia/A104-24-2007E.pdf

Christensen, Clayton. 1997. *The Innovator's Dilemma: When New Technologies Cause Great Firms to Fail.* Boston: Harvard Business School Press.

Commission on Intellectual Property Rights. 2002. *Integrating Intellectual Property Rights and Development,* ed. John H. Barton, Daniel Alexander, Carlos Correa, et al. London: Commission on Intellectual Property Rights.

Costa-Font, Montserrat, José M. Gil, and W. Bruce Traill. 2008. "Consumer Acceptance, Valuation of and Attitudes towards Genetically Modified Food: Review and Implications for Food Policy." *Food Policy* 33 (2): 99–111. http://dx.doi.org/10.1016/j.foodpol.2007.07.002.

Costanigro, Marco, Stephan Kroll, Dawn Thilmany, et al. 2014. "Is It Love for Local/Organic or Hate for Conventional? Asymmetric Effects of Information and Taste on Label Preferences in an Experimental Auction." *Food Quality and Preference* 31: 94–105. http://dx.doi.org/10.1016/j.foodqual.2013.08.008.

Costanigro, Marco, Dawn Thilmany McFadden, Stephan Kroll, et al. 2011. "An In-store Valuation of Local and Organic Apples: The Role of Social Desirability." *Agribusiness* 27 (4): 465–77. http://dx.doi.org/10.1002/agr.20281.

Davis, Joshua. 2003. "Come to LeBow Country." *Wired*, February 2003.

Graff, Gregory D., Susan Cullen, Kent Bradford, et al. 2003. "The Public-Private Structure of Intellectual Property Ownership in Agricultural Biotechnology." *Nature Biotechnology* 21 (9): 989–95. http://dx.doi.org/10.1038/nbt0903-989.

Graff, Gregory D., Gal Hochman, and David Zilberman. 2009. "The Political Economy of Agricultural Biotechnology Policies." *AgBioForum* 12 (1): 34–46.

Graff, Gregory D., and David Zilberman. 2001. "An Intellectual Property Clearinghouse for Agricultural Biotechnology." *Nature Biotechnology* 19 (12): 1179–80. http://dx.doi.org/10.1038/nbt1201-1179.

–. 2007. "The Political Economy of Intellectual Property: Reexamining European Policy on Plant Biotechnology." In *Agricultural Biotechnology and Intellectual Property: Seeds of Change*, ed. J. Kesan, 244–67. Wallingford, UK: CAB International. http://dx.doi.org/10.1079/9781845932015.0244.

–. 2013. "The Research, Development, Commercialization, and Adoption of Drought and Stress-Tolerant Crops." In *Crop Improvement under Adverse Conditions*, ed. N. Tuteja and S.S. Gill, 1–33. New York: Springer. http://dx.doi.org/10.1007/978-1-4614-4633-0_1.

Graff, Gregory D., David Zilberman, and Alan B. Bennett. 2009. "The Contraction of Agbiotech Product Quality Innovation." *Nature Biotechnology* 27 (8): 702–4. http://dx.doi.org/10.1038/nbt0809-702.

–. 2010. "The Commercialization of Biotechnology Traits." *Plant Science* 179 (6): 635–44. http://dx.doi.org/10.1016/j.plantsci.2010.08.001.

Grossman, Gene M., and Elhanan Helpman. 2001. *Special Interest Politics*. Boston: MIT Press.

Heiman, Amir, and Oded Lowengart. 2008. "The Effect of Information about Health Hazards on Demand for Frequently Purchased Commodities." *International Journal of Research in Marketing* 25 (4): 310–18. http://dx.doi.org/10.1016/j.ijresmar.2008.07.002.

Heller, Michael A. 1998. "The Tragedy of the Anticommons: Property in the Transition from Marx to Markets." *Harvard Law Review* 111 (3): 621–88. http://dx.doi.org/10.2307/1342203.

Heller, Michael A., and Rebecca S. Eisenberg. 1998. "Can Patents Deter Innovation? The Anticommons in Biomedical Research." *Science* 280 (5364): 698–701. http://dx.doi.org/10.1126/science.280.5364.698.

Hendrickson, Chris T., Lester B. Lave, H. Scott Matthews. 2006. *Environmental Life Cycle Assessment of Goods and Services: An Input-Output Approach*. Washington, DC: Resources for the Future Press.

Hinxton Group. 2010. "Consensus Statement on Policies and Practices Governing Data and Materials Sharing and Intellectual Property in Stem Cell Science." October 31–November 2, University of Manchester, Manchester, UK.

–. 2012. "Consensus Statement on Data and Materials Sharing and Intellectual Property in Pluripotent Stem Cell Science in Japan and China." January 29–31, Kobe, Japan.

House, L., J. Lusk, S. Jaeger, et al. 2004. "Objective and Subjective Knowledge: Impacts on Consumer Demand for Genetically Modified Foods in the United States and the European Union." *AgBioForum* 7 (3): 113–23.

Jefferson, Osmat A., Deniz Köllhofer, Thomas H. Ehrich, et al. 2013. "Transparency Tools in Gene Patenting for Informing Policy and Practice." *Nature Biotechnology* 31 (12): 1086–93. http://dx.doi.org/10.1038/nbt.2755.

Just, Richard E., Darrell L. Hueth, and Andrew Schmitz. 2008. *Applied Welfare Economics and Public Policy.* Cheltenham, UK: Edward Elgar.

Kathage, Jonas, and Matin Qaim. 2011. *Are the Economic Benefits of Bt Cotton Sustainable? Evidence from Indian Panel Data.* Courant Research Centre: Poverty, Equity and Growth, Discussion Paper No. 80. Göttingen: Georg-August-Universität Göttingen.

Kitabu, Gerald. 2014. "Here Comes Drought-Tolerant Hybrid Maize Varieties." *IPP Media,* January 17. http://www.ippmedia.com/frontend/?l=63832.

Kloppenburg, Jack R. 1988. *First the Seed: The Political Economy of Plant Biotechnology.* Cambridge: Cambridge University Press.

Lusk, Jayson L., Lisa O. House, Carlotta Valli, et al. 2004. "Effect of Information about Benefits of Biotechnology on Consumer Acceptance of Genetically Modified Food: Evidence from Experimental Auctions in the United States, England, and France." *European Review of Agriculture Economics* 31 (2): 179–204. http://dx.doi.org/10.1093/erae/31.2.179.

Lynas, Mark. 2013. Lecture to Oxford Farming Conference, January 3, 2013, Oxford, UK.

Marden, Emily. 2003. "Risk and Regulation: U.S. Regulatory Policy on Genetically Modified Food and Agriculture." *Boston College Law Review* 44 (3): 733–88.

Merges, Robert P. 1996. "Contracting into Liability Rules: Intellectual Property Rights and Collective Rights Organizations." *California Law Review* 84 (5): 1293–1393. http://dx.doi.org/10.2307/3480996.

Monsanto. 2014. "Roundup Ready Soybean Patent Expiration." Monsanto, http://www.monsanto.com/newsviews/pages/roundup-ready-patent-expiration.aspx.

National Institute of Food and Agriculture. 2014. "Grants – Minor Crop Pest Management Program Interregional Research Project #4 (IR-4)." United States Department of Agriculture, National Institute of Food and Agriculture.

National Research Council. 2010. *Impact of Genetically Engineered Crops on Farm Sustainability in the United States.* Washington, DC: National Academies Press.

Non-GMO Project. 2014. "Product Verification." Non-GMO Project, http://www.nongmoproject.org/product-verification/.

O'Connor, Sean, Gregory Graff, and David Winickoff. 2010. *Legal Context of University Intellectual Property and Technology Transfer.* Washington, DC: National Academy of Science, Science Technology Economics and Policy (STEP) Board.

Peltzman, Sam. 1974. *Regulation of Pharmaceutical Innovation: The 1962 Amendments.* Washington, DC: American Enterprise Institute for Public Policy Research.

Pharmaceutical Research and Manufacturers of America. 2013. "Intellectual Property Protections Are Vital to Continuing Innovation in the Biopharmaceutical Industry." Pharmaceutical Research and Manufacturers of America, http://www.phrma.org/innovation/intellectual-property.

Plomer, Aurora. 2013. "The Human Rights Paradox: Intellectual Property Rights and Rights of Access to Science." *Human Rights Quarterly* 35 (1): 143–75. http://dx.doi.org/10.1353/hrq.2013.0015.

Potrykus, Ingo. 2010. "Regulation Must Be Revolutionized." *Nature* 466 (7306): 561.

Reichman, Jerome H., and Catherine Hasenzahl. 2003. "Non-Voluntary Licensing of Patented Inventions: Historical Perspective, Legal Framework under TRIPS, and an Overview of the Practice in Canada and the USA." UNCTAD-ICTSD Project on IPRs and Sustainable Development, Issue Paper No. 5. Geneva: United Nations Conference on Trade and Development (UNCTAD) and International Centre for Trade and Sustainable Development (ICTSD).

Roe, Brian, Alan S. Levy, and Brenda M. Derby. 1999. "The Impact of Health Claims on Consumer Search and Product Evaluation Outcomes: Results from FDA Experimental Data." *Journal of Public Policy and Marketing* 18 (1): 89–105.

Romeis, Jörg, Detlef Bartsch, Franz Bigler, et al. 2008. "Assessment of Risk of Insect-Resistant Transgenic Crops to Nontarget Arthropods." *Nature Biotechnology* 26 (2): 203–8. http://dx.doi.org/10.1038/nbt1381.

Seufert, Verena, Navin Ramankutty, and Jonathan A. Foley. 2012. "Comparing the Yields of Organic and Conventional Agriculture." *Nature* 485 (7397): 229–32. http://dx.doi.org/10.1038/nature11069.

Sexton, Steven, David Zilberman, Deepak Rajagopal, et al. 2009. "Role of Bio-technology in a Sustainable Biofuel Future." *AgBioForum* 12 (1): 130–40.

Sharma, Kiran K. 2010. "Platform for Translational Research on Transgenic Crops (PTTC)." Presented at the Fifth Asian Biotechnology and Development Conference, Sri Lanka, December 15–18.

Smith-Spangler, Crystal, Margaret L. Brandeau, Grace E. Hunter, et al. 2012. "Are Organic Foods Safer or Healthier than Conventional Alternatives? A Systematic Review." *Annals of Internal Medicine* 157 (5): 348–66. http://dx.doi.org/10.7326/0003-4819-157-5-201209040-00007.

Stigler, George J. 1971. "The Theory of Economic Regulation." *Bell Journal of Economics and Management Science* 2 (1): 3–21. http://dx.doi.org/10.2307/3003160.

Sullins, Martha, Dawn Thilmany McFadden, Dominique Songa, et al. 2012. *Colorado Attitudes about Agriculture and Food.* Fort Collins: Colorado State University Extension.

Varshney, R.K., W. Chen, Y. Li, et al. 2011. "Draft Genome Sequence of Pigeonpea (*Cajanus cajan*), an Orphan Legume Crop of Resource-Poor Farmers." *Nature Biotechnology* 30 (1): 83–89. http://dx.doi.org/10.1038/nbt.2022.

Wu, Felicia. 2004. "Explaining Public Resistance to Genetically Modified Corn: An Analysis of the Distribution of Benefits and Risks." *Risk Analysis* 24 (3): 715–26. http://dx.doi.org/10.1111/j.0272-4332.2004.00470.x.

4

Stealth Seeds

Bioproperty, Biosafety, Biopolitics

RONALD J. HERRING

Monsanto, Terminators, and the Mud

In July 2003, before a meeting in Palakkad district, South India, to memorialize the peasant leader Keraleeyan, agrarian activists told me and each other of threats from "the terminator." I explained to colleagues that the "Monsanto/terminator/suicide-seed" narrative about *Bt* cotton tests was a product of a Canadian website, NGOs, and instrumental political dramaturgy, not reality. Politely, no one corrected me. At the meeting, a prominent public intellectual, P.A. Vasudevan, said that the current stage of historic agrarian struggles for which Kerala is justifiably well known was only for "the mud"; world agriculture would be controlled by Monsanto and Cargill, through biotechnology. Popular forces had learned how to struggle against and defeat the landlords, their *goondas*, and police, but they did not know how to fight globalization.[1] The emergence of a new farmer organization in a district already intensively organized reflects his analysis: the Deshiya Karshaka Samrakshana Samithi (National Agriculturalist Protection Committee, DKSS) was formed to protect farmers from globalization, one prominent manifestation of which was Monsanto and its terminator technology (DKSS 2003, 2).

In this diagnosis, the DKSS joined a loose national movement linking external threats to agriculture to multinational corporations and biotechnology. Ownership of a patent on terminator technology was (falsely) attributed to Monsanto, which was (even more falsely) said to have unleashed this bio-cultural abomination on India through field trials of *Bt* cotton. Terminator technology would in theory permit engineering of plants that could not produce viable seeds, causing a biological dependence of farmers on firms beyond that of commercial arrangements.[2] Traditional practices of "self-organizing" agriculture would be replaced by a dependency and cash nexus. This construction – linking multinational capital and globalization to the cultural abomination of "suicide seeds" – created a powerful political narrative. Paired with Dow Chemical Company, which "brought us Bhopal and Vietnam," Monsanto was said to be planning to "unleash genetic catastrophes" (Research Foundation for Science, Technology and Ecology 2003). The terminator imaginary persisted in Indian public discourse around biotechnology, demonstrating great power, pervasive reach, and persistence in the face of disconfirming evidence. Biopolitics centred on this ideational construction cleaved coalitions seeking social justice and betterment of farmers.

Suicides by debt-ridden farmers – most notably in Warangal district, Andhra Pradesh – were linked explicitly by activists to globalization of agriculture and new seed technologies (Stone 2002; Shiva et al. 2000, 64–110; Jha 2001).[3] Dependence of farmers on hybrid seeds of multinationals – rhetorically branded "seeds of death" or "suicide seeds" – linked field trials of transgenic cotton in 1998 to the opening wedge of terminator technology in India. The book *Seeds of Suicide* was "dedicated to the farmers of India who committed suicide" (Shiva et al. 2000). Chapter 1 of Vandana Shiva's *Biopiracy: The Plunder of Nature and Knowledge* (1997) is titled "Piracy through Patents." Dr. Shiva's overriding concern with biotechnology is "the control of agriculture by multinational corporations" (Shiva 1997, 91). Activists burning field crops of *Bt* cotton trials called their movement "Operation Cremate Monsanto." In response, terminator seeds were specifically banned by the Indian government in 1998, but the movement continued (Herring 2005).

Monsanto's representative in India publicly refuted charges of suicide seeds: "Since the so-called terminator gene does not exist today in any plant in any country in the world, the question of its involvement in the field trials currently on in India does not arise" (Dow Jones 1998). B.R. Barwale, the chairman of Mahyco, Monsanto's Indian business partner, emphasized that the seeds being tested had been approved for trials by the national Department of Biotechnology and have "nothing to do with the so-called terminator genes" (Mistry 1998). Monsanto's marketing director for India argued that the farmers' suicides had nothing at all to do with Monsanto's *Bt* seeds (which were not even on the market), but ironically might have been prevented by its technology (Mistry 1998). With transgenic cotton, he said, farmers would have had less debt from pesticide purchase and less crop loss to bollworms – less poverty, fewer suicides. More obviously, since the hybrid transgenics under testing had been backcrossed into local cultivars, there were at least six generations of *Bt* cotton in India at the time of the field tests, a clear indication that no terminator genes were in the cotton (Ghosh 2001, 11). If anti-*Bt*, pro-farmer organizers were interested, a countervailing discourse was readily available – and a high-profile one at that.[4]

That transgenics that arrived in India through Monsanto yoked biotechnology to multinational capital. Had *Bt* cotton come through the Indian private or public sector – as is now unfolding with both Chinese and Indian varieties – biopolitics would have taken a different turn. In the event, the technology was inseparable from the property, from multinational capital, and from pressures on agriculture from globalization.

Transgenic Cotton in India: From Robin Hood to Cottage Industry

The vicissitudes of cotton farming in India are extreme. Yields are among the lowest in the world, the area under cotton the highest (James 2002). A bitter irony in the farmer suicides is that insecticides unable to protect crops – because of either insect resistance or dilution and adulteration – were sufficiently strong to kill farmers ruined by debts incurred to purchase pesticides. Sharad Joshi, leader of India's largest farmer organization, Shetkari Sanghatana, illustrated the crisis through a single tragedy: "It [2001] was a year of miseries for the cotton growers of Maharashtra. Neelkanth Mankar, a cotton grower in Yavatmal district, unable to face creditors, committed suicide" (Joshi 2001).

In the neighbouring state of Gujarat, however, Joshi (2001) noted:

> Through a lucky stroke a nondescript seed company managed to play Robin Hood and smuggle into Gujarat one line of anti-bollworm gene. For three years nobody noticed the difference and then came the massive bollworm rampage of 2001.

There was no way to distinguish transgenic lint or seeds from their appearance, but fields indicated the difference (Joshi 2001):

> Gujarat saw all its traditional hybrid cotton crop standing devastated, side-by-side the *Bt*-gene crops standing resplendent in their glorious bounty.

The Government was upset and ordered destruction and burning of the bountiful crop.[5]

Neither the "Cremate Monsanto" movement nor the government bio-safety regulators had noticed the transgenic cotton. Monsanto's partner Mahyco did, and complained to the Genetic Engineering Approval Committee (GEAC) in Delhi. The seeds had been sold as Navbharat 151. They were originally detected in Gandhinagar district of Gujarat in six locations. Press reports typically said that the extent of coverage was over "10,000 hectares" (or sometimes "10,000 acres") in extent; this number was used in Parliament, and has crept into academic accounts. Both suspiciously parallel estimates are groundless. Precisely because these were underground seeds, no one knows exactly the extent or location of plantings. The GEAC investigated during the last week of September, and found Mahyco's charges to be true. The cotton contained the Cry1Ac gene in the construct of Monsanto. The head of Navbharat, Dr. D.B. Desai, was summoned to Delhi for the October 9, 2001, meeting of GEAC to explain the evident violation of biosafety regulations: no transgenic crop could be planted without officially sanctioned tests and final approval by the GEAC. Dr. Desai did not appear; his counsel argued that Navbharat did not know there was a *Bt* gene in its seeds. It was sold as a hybrid registered by the government of Gujarat. On October 12, 2001, the GEAC met again and ordered the Gujarat government to act. Here regulation encountered the light-switch problem: a switch is thrown but it is not connected to anything. Gujarat state had not set up a biosafety committee, as all states were mandated to do; the GEAC itself has no police powers.[6] A case was registered against Dr. Desai for violation of the *Environment (Protection) Act, 1986* and *Rules for the Manufacture/ Use/Import/Export and Storage of Hazardous Microorganisms, Genetically Engineered Organisms or Cells* (Rules of 1989) that regulate transgenic organisms.

The GEAC ordered not only burning of the crop, as the farmers' organization notably scorned in their resistance ("we will burn with our crops in our fields"), but also mandated: (1) a public warning in regional newspapers; (2) retrieval and destruction of seeds from farmers' houses and ginning mills; (3) collection of lint, which was to be stored in steel containers and transported to the Central Institute of Cotton Research in Nagpur for testing; (4) procurement of all unharvested crops from farmers; (5) uprooting and burning of the standing crop; and (6) measures to sanitize the fields.

These orders were not carried out, marking an unambiguous political victory for farmers over regulators. Appropriately enough, Gujarat's decision to do nothing to enforce the order was announced in Delhi by Union Textiles Minister Kashiram Rana immediately after a meeting with the chief minister of Gujarat, Narendra Modi. Delhi has a deep national interest in cotton production. Rana could see nothing wrong with the controversial seeds; he reasoned that since the *Bt* seeds reduced pesticide use and were favoured by farmers, opposition must be coming from the pesticide lobby. Gujarat state's minister of agriculture, Purshottam Rutala, made the same argument: who other than the pesticide lobby had an interest in depriving farmers of a beneficial technology? The consensus, across state and national governments, and eventually the GEAC itself, articulated by the secretary of the Department of Biotechnology, Manju Sharma, was that the "interests of farmers" would not be harmed. Union Agriculture Minister Ajit Singh said that delay in approval of the *Bt* seeds would be "inexcusable," given the damage bollworms were doing to Indian cotton. The result was that farmers would continue to grow, save, and breed transgenic cotton despite clear evidence of violation of India's environmental protection laws and biosafety procedures.[7]

The failure of the GEAC order indicated that the biosafety regime was out of the hands of regulatory authorities and scientists in 2001 and in the hands of politicians and organized farmers. The central government provided little political support for the hard line originally adopted by the GEAC. Regulation in a federal political system necessitates balancing central and provincial powers. Agriculture is constitutionally a state matter in India, and Gujarat's position was clear. Likewise, the government of the neighbouring state, Maharashtra, was actively pressing for immediate approval of transgenic cotton; Maharashtra is the geographic base of the Shetkari Sanghatana. In the State Legislative Council, Maharashtra Agriculture Minister Rohidas Patil announced on December 11, 2001, that the state would make *Bt* cotton seeds available to farmers from January 1, 2002, onward – three months before the GEAC in Delhi approved the technology.[8] On March 25, 2002, farmer representatives led by Sharad Joshi, a member of the Kisan [agriculturalist] Coordination Committee (KCC), threatened to launch a civil disobedience movement if *Bt* cotton was not approved by Delhi. KCC representatives from cotton-growing states across India – Gujarat, Maharashtra, Punjab, and Andhra Pradesh – rallied for immediate approval, and threatened to cultivate transgenic varieties whether or not the government approved (Herring 2005).[9] The following day, March 26, the GEAC approved three varieties of the Mahyco-Monsanto *Bt* cotton, making India the sixteenth nation in the world to certify a genetically engineered plant for commercialization, albeit provisionally and in the face of fierce opposition.

Although the officially sanctioned varieties – Bollgard MECH-12, -162, and -184 – are much more expensive than other hybrid seeds, they sold well and the technology has been licensed to other seed firms. Competition from the stealth seeds and their progeny has been vigorous, however. On August 16, 2004, Union Agriculture Minister Sharad Pawar stated in Parliament that the underground seed market was flourishing and alarming. No one knows how many "Robin Hoods" – in Sharad Joshi's analogy – are active in rural India, but it is clear that a cottage industry of transgenic pocket breeding has grown up around descendants of the original Navbharat 151 seeds.

There is a rough sequence to this evolution. After Navbharat 151 was banned in 2001, it became scarce though not impossible to find; farmers sought out its parent lines for breeding.[10] Some saved their seeds after ginning and sold or exchanged or replanted the F2 generation of Navbharat 151, which was no longer legally available in the market. There is deep irony in this spread of the vigorous offspring of the "suicide seeds." These seeds are called "loose seeds," straight from the ginning mill, unpackaged and unbranded.[11] They may express less *Bt* endotoxin, but, according to farmers in Gujarat, offer reasonable protection at a very low price.[12]

Nevertheless, many farmers worry about hybrid vigour in cotton and distrust F2 seeds. Given the high cost of official seeds and the scarcity of the very effective Navbharat 151 farmers themselves began breeding new transgenic hybrid varieties. They use Navbharat 151 seeds for the male contribution and a local variety especially well suited to their agronomic conditions for the female. From this process, a new Gujarati word has been hybridized: *Navbharat variants.* There are uncounted, branded, and packaged *Bt* variants in circulation: Luxmi, Kavach, Viraat, Sarathi, Vaman, Agni, Rakshak, Maharakshak, Kranti, the generic Kurnool *Bt,* and simply "151," playing on the original Navbharat 151 variety, among many others. These locally backcrossed hybrids made by farmers are sold by local merchants, who sometimes guarantee the seeds, to distinguish them from the many spurious seeds claiming *Bt* status in the market. To indicate transgenic character semi-covertly, some variants have printed on the package "BesT Cotton Seed." There are also farmer-to-farmer transactions of modified and crossed transgenic seeds with no name.[13] The decision matrix of farmers facing this volatile seed market is complex, as there is great agronomic and cost variation, but farmers in Gujarat have largely naturalized stealth *Bt* as part of their time-tested decision matrix (Roy, Herring, and Geisler 2007).

The tension between official seeds and stealth seeds is dynamic and in continuous flux as seed firms jockey for position and new varieties are approved – or decertified for certain areas, as were two Mahyco-Monsanto Biotech Limited (MMBL) hybrid varieties in Andhra Pradesh in 2005. When the GEAC refused approval for Mahyco-Monsanto's new variety, MECH-915, for growing in North India, an advertisement soon appeared in a prominent Hindi daily advertising farmer-grown transgenic seeds available by calling the cell phone of one Piyush Patel. Sonu Jain, relating the story in the *Indian Express* (Delhi) of April 20, 2002, quotes Patel as saying: "If I live in Gujarat and go to Shimla, I will not die, so the same way these seeds developed in Gujarat will grow." Patel's *Bt* seeds sold at 555 rupees per packet of five hundred grams, less than a third the officially approved Mahyco-Monsanto seeds' price. By June 2005, I found that the range of locally hybridized transgenic cotton cultivars in Gujarat sold for 250–700 rupees per packet (roughly enough to plant one acre); F2 transgenic seeds were selling for 10 rupees for a packet of the same weight. Jayaraman (2004b) cites "industry sources" as estimating that over half of the transgenic cotton in India comes from unapproved varieties; my discussions with Gujarati seed producers suggest a much higher figure for that state. Data from Navbharat Seeds (personal communication, October 2005) indicate that on an all-India basis, about 34 percent of the cotton seed packets sold are transgenic, of which 9 percent are legal and 25 percent stealth. Yet these estimates apply only to packaged and branded stealth seeds, not to loose seeds. The ratio is highest in the North Zone (Punjab, Haryana, and Rajasthan): 107,000 packets of legal transgenic seeds to 1,170,000 illegal packets, together accounting for about a third of cotton acreage.

Dr. R.P. Sharma of the GEAC believes this to be a temporary phenomenon. Farmers, he says, will eventually choose *Bt* cultivars from trusted seed companies and abandon the stealth transgenics; the current state of affairs simply reflects the fact that "scarcity breeds corruption" (personal communication, New Delhi, June 27, 2005). Scarcity was induced by banning the extremely effective Navbharat 151.

Where did stealth *Bt* originate? Standard accounts have emphasized Gujarat and the Robin Hood construction of Sharad Joshi, but the seeds were available in other locations as well. In early November 2001, 460 acres of seed farms were found to be producing transgenic cotton in the Kurnool and Mahabubnagar districts of Andhra Pradesh. GEAC chairman A.M. Gokhale claimed that Navbharat was selling transgenic cotton under the brand names *Vijay, Digvijay,* and *Jay* in the state. It was widely believed that other strands of parent seeds were growing in the Punjab and Maharashtra.[14] Dr. N.P. Mehta, the breeder who developed Navbharat 151 has stressed serendipity: his breeding strategy was directed toward early flowering for a crop rotation with sugar cane in Surat; one of his scouts came across an unusually bollworm-resistant cultivar, which proved vigorous on crossing, producing Navbharat 151 (personal communication, Ahmedabad, June 20, 2005; Mehta 2005). Both MMBL and the GEAC believe that the transgene was appropriated from MMBL stock, possibly from the field trials mandated by the GEAC. Whatever the primary source, it is now clear that sequestering of *Bt* technology in either the regulatory or property sense has not proven administratively or politically feasible.

Farmers in Gujarat have embraced stealth seeds. This outcome is not in the interest of seekers of innovator rents through state protection of intellectual property – in this case both Monsanto and Navbharat. The long development costs and time, mandated by a biosafety regime that is becoming global, are estimated to be about US$8 million for MMBL; these costs put official seeds at a price disadvantage compared with illicit seeds. In Indian law, there is no restriction on farmer-to-farmer exchange of seeds. Farmer-generated transgenic hybrids that I have seen refer on the package to the *Indian Seeds Act, 1966,* section 24, which protects this right explicitly. Lacking capacity to use patent protection on intellectual property, Mahyco-Monsanto favours strict regulation. In these circumstances, regulatory restriction of official seed varieties confers property-like rights unavailable through patents. If only its seeds are legal, licensing of the technology confers short-term monopoly rents on MMBL. Advocates for Navbharat Seeds accuse the GEAC of exactly this: market rigging through biosafety regulation creates de facto bioproperty rights.

Bt cottons have been in the field too short a time for definitive assessment of either biological or economic success across so varied an agroecology as India; results vary with seasonal variations of pests, weather, and local agronomics (Rao 2004b). More systematic data are being compiled, much of it rejected by opponents for being tainted by Mahyco-Monsanto sponsorship, and none long-term enough for robust conclusions.[15] But unless one thinks farmers irrational, there is strong evidence of *Bt* technology's effects in India being similar to those in China, where both the Monsanto and public sector versions of *Bt* cotton have been adopted rapidly by small farmers (James 2002; Pray et al. 2002). Farmers have done this for higher yields, less pesticide application against bollworms, and higher profits.

Opponents continue to proclaim "the disaster wrought by *Bt* cotton" in India.[16] This construction is empirically groundless but strategically partisan. Opponents somehow fail to distinguish *Bt* technology from specific cultivars. The reason that there are between two hundred and three hundred cultivars of cotton grown in India is that cultivars have specific agronomic characteristics: no single variety will do well in all places in all seasons. Farmers know this; many who claim to represent them do not.

There exists no credible evidence of the "failure" of *Bt* cotton technology in India. There is great variation in performance of cotton cultivars, both *Bt* and non-*Bt*. Reasons for such variation are not always discernible, either by farmers or by researchers, since there are many unmeasured variables in complex interactions – local climate, soil chemistry, pest variance, water timing, nutrients. Second, spurious seeds are pervasive: some varieties sold as *Bt* are not; some farmers honestly but mistakenly believe their *Bt* crop has failed. Third, there are demands for financial compensation from Mahyco-Monsanto and the government for *Bt* crop failure; there is material incentive to claim poor results. Fourth, the MMBL varieties are clearly not the best germplasm for insertion of the *Bt* gene: many farmers seem to prefer Navbharat and other varieties, legal and illegal. New firms are vigorously entering the market as licensees of MMBL's technology, but with different cultivars. Finally, and most important, none of the claims of failure compare two isogenic varieties, one with and one without the *Bt* gene, to ensure control of varietal characteristics (Naik et al. 2005). Rather, all disadvantageous variance over time and space – which is extreme in India – is attributed to the *Bt* gene, constructing a biological absurdity. The *Bt* gene codes for a single protein, the Cry1Ac; there is no reason for production of that protein – lethal to Lepidoptera – to cause staples to shorten or leaves to wilt.[17]

Perhaps the most carefully controlled study to date is of the variety MECH-162 (Bambawale et al. 2004) – compared with the isogenic non-*Bt* MECH-162 and a conventional hybrid. This study used a participatory field trial to test meaningfully paired varieties with and without integrated pest management. Consistent with other studies, *Bt* plants required half the sprayings of other plants, experienced less bollworm infestation, and, somewhat surprisingly, reduced attacks of sucking pests and two natural enemies of cotton. With integrated pest management, the *Bt* variety recorded a yield of 7.1 quintals per hectare (q/ha) and a net return of 10,507 rupees per hectare.

Damage to fruiting bodies was much less with *Bt* plants, which would account for the premium some *Bt* farmers receive for their lint in the market. The authors concluded: "*Bt* Mech-162 used in an IPM [integrated pest management] mode resulted in highest yields and economic gains to the farmers; pesticide consumption was also reduced" (Bambawale et al. 2004, 1633). *Bt* technology and improved agro-ecological practices each contributed to superior outcomes in this controlled study.

Indian farmers are experimenting widely with *Bt* cottons, both official and unofficial; both categories are multiplying rapidly. Neither duped nor passive puppets of multinational monopolists, cotton farmers continue the primordial struggle of agriculture against insects, with a new tool. Their techniques continue traditions of seed saving, seed exchange, and seed experimentation (Gupta and Chandak 2005; Roy, Herring, and Geisler 2007). But then what of the elaborate biosafety regime that is to prevent genetic anarchy? A commission on biosafety regulations headed by eminent agricultural scientist M.S. Swaminathan concluded in understated officialese: "Public regard and satisfaction for the regulatory systems currently in place are, to say the least, low" (Bagla 2004). Dr. Swaminathan himself was quoted in *Nature Biotechnology* (Jayaraman 2004a, 1334) as saying that "illegal proliferation of GM varieties must cease or else the bio-safety regulations will be rendered meaningless." Yet A.K. Dixit, director of agriculture for Gujarat, said: "It is impossible to control something at this large a scale. When we go to the fields, we become targets for trying to take away a beneficial technology from farmers" (SeedQuest 2003a).

Representing Farmers: Biopolitics and Credibility

Coalitions for the poor are difficult to conjure, harder to create and sustain. The experience of a single farmer, little noted at the time, proved diagnostic of the representational problems of the suicide-seed coalition.

At the beginning of Operation Cremate Monsanto in 1998, the Karnataka Rajya Raitha Sangha (KRRS) organized the burning of *Bt* cotton crops on two test plots in Raichur and Bellary districts of Karnataka state. The KRRS is a farmer organization specifically dedicated to protecting Indian farmers – and India – from globalization, personified by Monsanto (Herring 2005; Omvedt 2005). Cremated test plots received international attention; failure to cremate a third trial plot as planned, in Adur village, Haveri *taluq*, went virtually unnoticed. Yet this episode proved diagnostic of difficulties facing any coalition that seeks to be both pro-poor and anti-transgenic. The farmer who owned the plot, *Shri* Shankarikoppa Mahalingappa, was a member of the KRRS but was unconvinced by its leadership. He stressed in a conversation with me that he could not depend on the movement's explanation of the new technology, but had to see for himself. He asked for and received police protection for his test crop. *Shri* Mahalingappa had sequestered one hundred seeds from biosafety trials on his land; he planted them and found germination by ninety-five of the one hundred "suicide seeds." The terminator construction dissolved before the organic empiricism of a farmer concerned about pesticide costs and farm income. Mahalingappa concluded that "terminator" talk was "just propaganda."[18]

Mahalingappa's dissent illustrated not only the political problem of cultural distance of movement elites from farmers but also a different view of science. His epistemology was inductive and grounded rather than deductive and derivative. Against the notion that farmers were duped by corporate propaganda, he said: "No one's word can be taken; you have to see for yourself ... farmers must be convinced personally that a crop is beneficial; only the farmers can decide." His perspective is simply prudent farm management. Mahalingappa now buys Bollgard MECH-162 and finds it profitable despite high seed cost; he spends less on chemicals and gets higher yields because of reduced bollworm damage. Why does he not use the cheaper stealth seeds so popular in Gujarat? He trusts the quality of the official seeds and does not trust the F2 generation of hybrids. The additional cost of seeds is more than compensated by additional revenue: "It makes money for me." And what of biosafety? "The genes cannot be taken back," but he is not worried about any bad effects because he's seen none. The foliage seems not to harm insects other than bollworms, nor mammals; he could perceive no threat from the new seeds.

How could there be such a disjuncture between farmer experience and the position of public intellectuals and NGOs that claim to represent the farmer? Consider this typical press account of Operation Cremate Monsanto:

> "Farmers," it is said, led by Prof. Nanjundaswamy of the Karnataka Rajya Raitha Sangha, attacked the Monsanto seed farm near Malaldguda in Raichur district and destroyed the cotton crop in order to protect Indian farmers from the dreaded Terminator Gene. The campaign was repeated in Bellary district and other "farmers" in Andhra Pradesh did the same thing. Speaking in the name of all of them, Prof. Nanjundaswamy has vowed to repeat it in Maharashtra and Punjab.[19]

This press report notes the ambiguity in characterizing "farmers" active against transgenics. It is now clear that the technology was spreading, not terminating, because farmers were finding ways to obtain illegal seeds and eventually ways to breed transgenic hybrids themselves.

Despite persistent reports of catastrophic failure and movements to de-certify Mahyco/Monsanto varieties (successful in Andhra Pradesh in 2005), demand for official seeds continues to grow, at times exceeding supply, even at prices triple or more those of other hybrids. Sales of the MMBL Bollgard seeds increased sixfold over the previous year in 2004–05. More-over, twenty-one varieties produced by Indian firms that had sublicensed the *Bt* technology had been approved for planting as of June 2005. Approv-als of new *Bt* cultivars continue to come from the GEAC, including the Bollgard II technology. Farmers continue to scramble for *Bt* seeds but are price-conscious, often opting for stealth seeds over official seeds.[20]

Yet "the failure of *Bt* cotton" – in an agronomic and economic sense – continues to fill press reports. Refusal to believe that farmers might have some valid experience on which to base a preference for transgenics is diagnostic of representational problems in rural movements headed by metropolitan elites. This epistemological fundamentalism – privileging international public intellectuals over farmers' intelligence – creates a wedge in possible coalitions for the poor.[21] If one grants that Indian farm-ers and seed firms might be commercially rational, the continuing spread of the Cry1Ac technology must indicate biological and micro-economic success. When I presented this argument to activists against biotechnology in Palakkad district, the response was that farmers had been duped or co-erced, falling into the trap of the monopolist Monsanto. Vandana Shiva and colleagues wrote in the conclusion to their 1999 article in the prominent Indian social science journal *Economic and Political Weekly* that "the pro-motion of genetic engineering by corporations like Monsanto can only be based on dictatorial, distorted and coercive methods."

If biological and economic failure constitute one strand of resistance to a technology many farmers find beneficial, a second strand is nationalism: biotechnology is unacceptable because of its foreign origin and implica-tions for neocolonial control. Responding to a BBC story that portrayed the farmers of Gujarat as clever pirates of Monsanto's intellectual property – as implied by Sharad Joshi's Robin Hood characterization – the Research Foundation for Science, Technology and Ecology (RFSTE, headed by Vandana Shiva) rebutted:

This rumour about piracy is initiated by Monsanto whose *Bt* cotton has totally failed throughout the length and breadth of the country and to divert attention of the public and policy makers from the failure of its genetically engineered seeds, Monsanto is trying to focus on the outstanding success as unjust and illegal of an indigenously bred cotton variety.[22]

"Indigenously bred," perhaps, but equally transgenic – with the Monsanto Cry1Ac gene.[23] Charges of genetic pollution and the saving construction of indigenous breeding provide an exit from the cul-de-sac of the suicide-seed narrative. Just as the "Monsanto terminator" construction came from a website in Canada, the revisionist response of anti-transgenic forces bears a striking resemblance to Canadian websites defending Percy Schmeiser. The construction of Canada's most famous biopirate (in Monsanto's view) or victim of biological pollution (in the opposition's view) deploys the defence of gene flow. Schmeiser's defence in the case he lost repeatedly in the lower courts and finally in the Supreme Court of Canada in June 2004, is that he did not illegally appropriate intellectual property for transgenic canola seed production on his farm, but suffered pollution of his fields by Monsanto's plants.[24] The revisionist reconstruction of RFSTE valorizes Gujarati farmers' "indigenous" breeding, but the major trait responsible for success – resistance to bollworms through the Cry1Ac endotoxin – is reconstructed as incidental and accidental, a result of biological pollution from Monsanto.

The *Bt* controversy began with a nationalist (and Gandhian) theme of resisting foreign threats to India (Herring 2005). Attacks on Monsanto continue, but *Bt* technology itself has been naturalized. The leader of the farmer organization Khedut Samaj in Gujarat, Bipin Desai, charged that the "failing" – and foreign – *Bt* technology has been approved by the government, whereas the successful home-grown *Bt* variety (Navbharat 151), has not. The organization's vice president, Labshankar Upadhyay, added: "The BJP talked about *Swadeshi* [self-reliance]. But it promotes a foreign company at the cost of an Indian firm. And we [farmers] stand to lose." The RFSTE construction enables a nationalist and populist attack on the biosafety regime: Delhi officially allowed the import and testing of, and then certified, a foreign incarnation of *Bt* technology, but filed a court case against an indigenous plant breeder. The stealth seeds are held to be legitimate in a way that the government seeds – which are much more expensive – are not (Mehta 2005, 77–84, 108–15, 137–39).

Stealth seeds were decisive in turning Indian farmers into a pressure group for transgenic cotton. Farmers encountered transgenics through the mediation of rumour, misinformation, and contradictory official signals (Parmar and Visvanathan 2003). The suicide-seed narrative was dramaturgically engaging, but external political intermediation loses power over time unless both framing and objectives resonate with a threshold level of farmers' experience. The micro-economic and biological success of *Bt* technology outweighed the more indirect, distal, and hypothetical arguments about foreign control and dangerous genes: oppositional discourse outran agricultural interests. Framing is bounded, ultimately, by interests, even if loosely at any particular time.

The triumphalism of the industry relies on the success and acceptance of *Bt* technology but may be premature. For all the romanticization of local knowledge and the village *Volk*, it is not clear that sons of the soil always know best (Herring 2000). Farmers adopted insecticides not after considering the science and social externalities but rather from an interest calculation: to protect their crops. Farmers understand the dead-end nature of the pesticide alternative to *Bt* cotton; they are little concerned about biosafety. Uncertainties are externalized to society as a whole, and projected onto an elusive "biosafety regime."

Stealth Soy in Brazil

India's *Bt* cotton story suggests the need to rethink core assumptions in the standard narrative of seeds and states in developmentalist biotechnology. Is India an aberrant case?

Brazil parallels India's experience in three ways. First, the developmental state supported biotechnology. Looking outward to competition in the international economy, Brasília pressed for indigenous and collaborative development of biotechnology. In both nations, support for transgenics from the centre was challenged in the periphery, by forces in civil society. Second, despite considerable political conflict around the biosafety regime, regulations were rendered irrelevant by the capacity of farmers to find and breed transgenic seeds underground. Third, Monsanto provided a symbolic target and rallying point for opposition at the same time that farmers were finding ways to avoid Monsanto's property claims. Rather than monopoly power, Monsanto encountered not only competition from stealth seeds but the same difficulty in policing underground seeds faced by the regulatory state. Intellectual property rights fiercely debated in political and policy circles proved illusory when seeds went underground.

Brazil's economy depends on agricultural exports, of which soy is a major component. As competing nations approved transgenic soy much sooner,

Brazilian farmers faced cost disadvantages vis-à-vis farmers in Argentina and the United States – the other two of the top three soy exporters (Paarlberg 2001, 67–92; Poddar 2004, 24–52).[25] But global imperatives were contradictory in the case of soy. Global market segmentation created a niche for "GMO-free" soy when opposition to transgenics surfaced in Europe and Japan after 1997; the cost-of-production disadvantage Brazilian farmers faced might be offset by the premium for GMO-free soy (Paarlberg 2001, 67) – if exports could be reliably segregated and certified.

Introduction of transgenic crops to Brazil proceeded in what David Hathaway (2002) called "a vacuum in the exercise of authority." The state was divided against itself. The political struggle for legitimacy to rule on transgenics involved different levels of the court system, divisions within the federal government, and disputes between the states and Brasília.

Brazil's *Biosafety Law* was passed by the National Congress in 1994 and issued in January 1995. It granted authority over genetically engineered organisms – both pharmaceutical and agricultural – to a National Technical Biosafety Commission (Comissão Técnica Nacional de Biossegurança, or CTNBio), consisting of academics, a broad range of government ministries, from Health to Agriculture to Education to External Affairs, and representatives of industry and civil society. CTNBio was to be the federal agency responsible for the development and implementation of biosafety policies and ethical codes, and for the evaluation of risk and environmental threat. Brazil's 1990 *Consumer Defense Code* mandates labelling of all products to inform consumers of characteristics they have a right to know about. Transgenics are also regulated by the *Industrial Property Code* of 1996, which explicitly responded to new requirements of the World Trade Organization's *Agreement on Trade-Related Aspects of Intellectual Property Rights* (TRIPS Agreement), granting legal protection to inventions related to pharmaceuticals, food processes, and biotechnology (Niiler 1999; Buckly 2000; Paarlberg 2001, 73–74; Hathaway 2002; Fontes 2003).

CTNBio approved the commercial release of three varieties of Monsanto's Roundup Ready (RR) soybeans in September 1998, and the representative on the commission from the Institute for Consumer Defense (IDEC) resigned in protest. Both Greenpeace and IDEC filed legal appeals; in 1999, an injunction (*ação cautelar*) was issued by a federal judge, Antonio Prudente. Commercial cultivation of RR soybeans was legally banned, on the grounds that they had not been adequately tested for human health and environmental impacts. More fundamentally, the authority of CTNBio was challenged by a suit in 2000 that sought an injunction against decisions on transgenic-crop releases before the government formulated rules for assessing biosafety. A third decision, issued on February 14, 2002, in response to a suit brought by the federal Public Ministry, suspended all further field tests of "biopesticide" transgenics until Brazil's pesticide legislation was enforced. These decisions combined to produce a "judicial moratorium" on the commercial release of transgenic crops in Brazil.[26]

Opposition to the regularization of transgenic soy was led by Greenpeace Brazil and IDEC, later joined by a section of the Ministry of Environment – the Brazilian Institute for the Environment and Renewable Natural Resources (Instituto Brasileiro do Meio Ambiente e dos Recursos Naturais Renováveis, or IBAMA) – and the umbrella Campanha Por um Brasil Livre de Transgênicos (Campaign for a Transgenic-Free Brazil).[27] As in India, opposition targeted corporate globalization, and Monsanto in particular. At a "People's Tribunal" – a functional equivalent of the Beej Panchayat of India – Monsanto was convicted of producing seeds sold to Argentina in 1995, subsequently smuggled into Brazil, that "endangered the environment, biodiversity, human health, the country's agricultural genetic wealth, and the Brazilian economy" (Osava 2004; Associação Nacional de Biossegurança 2001).

The federal government held that uncertainty was discouraging private investment in biotechnology, both foreign and domestic, as well as development within the sophisticated public sector. In July 2000, six senior cabinet members of then-president Fernando Cardoso's government (including those for Environment and Health) signed a "manifesto" supporting both transgenic crops and the authority of the federal CTNBio. Brazil's National Academy of Science as well as the Brazilian Genetics Society expressed approval of transgenic crops (Poddar 2004, 30). In December 2000, legislation granted CTNBio more robust authority to circumvent appeals against its legal competence.

Much of the legal manoeuvring in elite circles was irrelevant to what was growing in the ground. A report issued in 2002 by Liberal Party parliamentary deputy Ronaldo Vasconcellos exposed widespread planting of glyphosate-resistant soy in Brazil despite the legal ban (Osava 2002). Seeds had been smuggled from Argentina, and perhaps other neighbours, since at least 1997, and were being reproduced and recrossed by Brazilian farmers. According to Vasconcellos's report, transgenic seeds accounted for up to two-thirds of the crop of Rio Grande do Sul, Brazil's southernmost state; there was also evidence of seeds spreading north.[28] Vasconcellos was suspicious of his own numbers and underscored the political consequences of stealth seeds for biotechnology data. He suggested that disseminating exaggerated figures had a political rationale: to present transgenic soy as a *situación de facto* (fait accompli), seemingly irreversible. Indeed, some opponents saw stealth seeds generally as a "contamination strategy."[29] As in Gujarat, the actual area planted to underground seeds in Brazil was unknown. Officially, Rio Grande do Sul maintained its own anti-transgenic policy and stopped most federally authorized field trials in the state. The state government claimed incapacity to halt widespread smuggling from Argentina, although by this time most seeds seem to have been indigenously produced.

The illegal-immigrant seeds themselves had not necessarily been purchased through legal channels. Monsanto was refused patent protection for Roundup Ready soybeans in Argentina in 1995. Argentina's largest seed firm, Nidera, purchased a former Monsanto partner, Asgrow Argentina, and thus acquired RR seeds, which were sold to farmers without technology fees. In 1999, the president of the National Seed Institute (INASE) of Argentina, Adelaida Harries, estimated that "25–30 percent of the soybean and wheat crops are sown with illegal seed, which farmers sell to their neighbours."[30] Nor could Monsanto count on enhanced chemical sales as its technology spread: Roundup (glyphosate) was out of patent and twenty companies competed with Monsanto for sales (Paarlberg 2001, 71–72). Opponents of transgenics in Rio Grande do Sul argued that control of the international border was a federal matter; the fault lay with Brasília, which favoured biotechnology. Other states – Santa Catarina, Mato Grosso do Sul, and Pará – passed state laws giving their newly created State Biosafety Commissions control over transgenics in anticipation of federal deregulation (Hathaway 2002). As in India, federalism made for a disjointed and indeterminate regulatory environment.

In January 2001, approximately 1,200 Brazilian farmers mobilized by the Landless Rural Workers' Movement (Movimento dos Trabalhadores Rurais sem Terra, or MST) stormed a Monsanto biotechnology plant in Rio Grande do Sul in protest against genetically modified food (Reuters 2001). This protest strategically coincided with the World Social Forum that had convened in Porto Alegre, countering the World Economic Forum simultaneously under way in Davos, Switzerland. Aided by antiglobalization protesters, farmers uprooted genetically engineered corn and soybeans at the experimental station and wrote on walls: "The seed of death" and "Monsanto is the end of farmers!" A local MST leader, Solet Campolete, explained that "these seeds trick farmers and create dependency on seeds produced by a big multinational."[31]

Although discursive framing of social movements paralleled that of India, there was also interest-driven opposition: stealth soy was a threat to Brazilian exports. Federations of agriculture in Mato Grosso do Sul – the country's leading soy-producing state – and Paraná demanded action against transgenic soy. Paraná began interdicting plantings in seventeen areas and ultimately attempted to blockade seeds by declaring the state a "GMO-free zone." Paraná had been conducting negotiations to export non-transgenic soy to China (*Actualización*, March 24, 2003, Issue 8). Both states' federations cited the preferences of the European Union and Japan for GMO-free food. The Paraná federation of agriculture's Technical Commission on Grains warned of "an enormous loss of revenues" because the national system is unable to certify "reliable separation" of transgenic soy (Osava 2002).

Opponents of Monsanto, globalization, and transgenics expected support from the federal level when the Workers' Party (Partido dos Trabalhadores, or PT) government took office on January 1, 2003. The party promulgated an antiglobalization agenda and endorsed the precautionary principle in environmental policy. On September 25, however, Presidential Decree MP131 authorized the planting and harvesting of transgenic soy for one year (Conroy 2003) – a de facto amnesty. In 2004, permission was extended for one year. Farmers were allowed to plant transgenic seeds saved from the last harvest, but not to sell them (Osava 2004; *Actualización*, September 25, 2003, Issue 3). In October 2004, permission for growing transgenics in the 2005 season was granted as well, subject to registration.[32]

The government had first sought to limit transgenic soy to one state, Rio Grande do Sul. When Governor Requião of the neighbouring state of Paraná declared both his state and Brazil's main grain port, Paranaguá, as GM-free zones, officials began stopping soy trucks at the border and turning away those that tested positive for glyphosate-resistant soy. This practice directly contradicted the presidential decree; it also created a problem of international jurisdiction, as countries such as Paraguay that permit transgenic soy found their shipments delayed or seized. On December 11, 2003, Brazil's Supreme Court unanimously ruled in favour of suspension of Paraná's law, while delaying a final ruling on constitutionality (Reuters 2003). The Liberal Front Party had brought the case, arguing that the state legislature had overstepped its jurisdiction. The perhaps unstable outcome is that there can legally be no "GM-free zone" in Brazil (SeedQuest 2003b). On March 26, another presidential provisional measure (later converted to Law 10688/03) established rules for the commercialization of the 2002/2003 soy crop, including transgenic soy. This measure gave amnesty to farmers who had illegally planted transgenic soy when the ban was in place. There are no official statistics, but there are estimates that up to 90 percent of soy in Rio Grande do Sul was transgenic (Bonalume Neto 2003; *Actualización*, December 5, 2003, Issue 6).

Significant sectors of civil society objected to post-facto regularization of stealth seeds in an economy that both consumes and exports soybeans and products on a large scale. The Campaign for a Transgenic-Free Brazil went so far as to brand President Lula, whom they saw as a proponent of the precautionary principle before his election, as a "crusader" for transgenics ("*un cruzado pro de-GM,*" *Actualización*, March 24, 2003, Issue 8). As in India, farmers' organizations were divided; the spread and replication of seeds underground clearly shows their desirability to many farmers. Some deputies in the Workers' Party supported what opponents, specifically Paulo Pimenta, called the agenda of big landlords and soya producers of Rio Grande Do Sul. Whether poor farmers opposed or supported transgenics is not established, but there is anecdotal evidence that small farmers have bought into the technology, which should in theory be scale-neutral.[33]

The national state is divided as well. The Agriculture Ministry has supported liberalization of transgenics, whereas the Environment Ministry, increasingly marginalized, has opposed it (*Actualización*, December 18, 2003, Issue 5). A new biosecurity law was proposed in October 2003; in February 2004, the amended draft bill was approved by the Chamber of Deputies. Delays and compromises reflected differences in the ruling coalition, particularly conflict between Agriculture Minister Roberto Rodrigues and Environment Minister Marina Silva, who has opposed deregulation in the absence of further studies (Bonalume Neto 2003; Hochstetler 2004, 17–19). Agricultural interests recognized these deep divisions and were reluctant to plant transgenic soy, fearing that actions by the courts, the environment minister, and state governments might prevent their marketing.[34]

Why were these seeds spreading? Urbashi Poddar (2004, 31) summarizes the advantages perceived by Brazilian farmers:

> Producers are able to control weeds with less tillage and limit their herbicide applications to just one spray over the top of the plants rather than adhering to a more expensive sequence of pre-emergent and post-emergent sprays ... used with conventional soybeans ... The Brazilian Association of Seed Producers estimates that farmers save approximately 20 percent on

their fertiliser costs using the genetically modified crop, "which also taxes their land less'" – with higher yields. In justifying use of illegal technology, some farmers responded that the seeds "serve as an instrument of economic leverage against the subsidy regime upheld by the United States and other … industrialized nations."

EMBRAPA (Empresa Brasileira de Pesquisa Agropecuária, or Brazilian Enterprise for Agricultural Research) scientist Maria José Sampaio gave me a simpler explanation: "They say it is like planting gold" (personal communication, 2003 and 2004).

Public distrust of transgenics targets opaque concentration of information in unaccountable hands. Although public institutions are being crafted to permit trials and supervise biosafety procedures, much research is under private auspices. Brazil, like India and China, has significant public sector resources for biotechnology. Paarlberg (2001, 70) notes that Brazil's intellectual property protections have encouraged not only foreign investment but also scientific entrepreneurs. Brazil's public sector research organization, EMBRAPA, has announced a new transgenic soy resistant to imidazolinone-based herbicides, similar to Monsanto's RR soy, which is resistant to herbicides of the glyphosate family. Testing for commercial release will take time, and imidazolinone is more expensive than glyphosate; nevertheless, the assumption of multinational monopoly of biotechnology is unsustainable. Such outcomes of public science undermine critiques of biotechnology that assume a high-technology monopoly of bioproperty (Reuters 2004b). Paarlberg (2001) notes that plant breeders at EMBRAPA have developed local varieties of soybean, maize, and cotton best suited to Brazilian conditions, and have used Brazil's new IP system to establish ownership of this valuable collection of improved germplasm. These scientists see strengthening of the research capacity of universities and public institutes as an antidote to monopoly by multinational firms (Paarlberg 2001, 70).[35]

The future of stealth seeds in Brazil remains in legal limbo. Farmers who grow them are in theory required by the federal government to sign an agreement to comply with terms for growing transgenic soy, but there are interpretive disputes as to what compliance means and how it can be enforced. Moreover, Monsanto claims intellectual property rights in the germplasm and continues to press its property claims, against uneven resistance by Brazil and its neighbours.[36]

How widespread are stealth seeds beyond Brazil and India? Anecdotal evidence is prolific, hard evidence scarce. Pray and Naseem (2007) note that "a recent survey of scientists in Latin America by ISNAR indicates ... that government scientists were using patented techniques extensively without knowing they were patented in some cases or not caring about it in others." Qaim and de Janvry (2003) find that if "black market" seeds for *Bt* cotton in Argentina were included in the calculus, most benefits of the technology would end up in farmers' pockets, not Monsanto's. Pray and Naseem (2007) also note that although China has patent laws, plant breeders' rights, and trademarks – in theory protecting both Monsanto and public sector *Bt* cotton varieties – and has stringent biosafety regulations, "most of the *Bt* cotton-seed that farmers planted was produced by farmers or small seed companies and government research institutes who did not apply for bio-safety committee approvals or pay royalties to the owners of varieties or genes."[37] *Bt* rice has reportedly appeared in the market in China without biosafety clearance. Bolivia and Paraguay have had the Brazil-Argentine experience with underground soy; Paraguay has explicitly acknowledged the fact.

Under what conditions will the phenomenon of stealth seeds emerge? Underground seeds have appeared in democratic and authoritarian systems, federal and centralized, strong state and weak state. Generalization is certainly premature. Certainly some common factors contribute: monopoly prices of alternative transgenic seeds, enabled by biosafety regulations more than market power; the structural power of agriculture as a sector and local power of farmers as political actors; permeable bureaucracies; regulatory confusion and delays; and weak institutions. These conditions, although widespread, tend to disappear in biotechnology discourse, both developmentalist and oppositional.

Reification of Seed and State

The ability of seeds to go underground through farmer stealth strategies undermines the surveillance of states and firms that is assumed in two divergent discourses: opponents of genetic engineering fear monopolization of property rights, whereas proponents assume biosafety regimes that can control nature and farmers – a Panopticonic state enforcing global protocols. Neither discourse is proving robust on the ground. Surveillance of nature is no mean task, either macro or nano.[38] Strong interests are at work. Farmers pursue stealth seeds because the technology is affordable and divisible; the genetic roulette they enable has not been a major concern.

Stealth seeds inject a new dimension of dissension in potential coalitions for the poor: farmers seek them out; NGOs seek to stop the technology. Brazil's PT government has been split, as have representatives of farmers and NGOs. In India, the loss of credibility resulting from the Monsanto-terminator-suicide-seeds campaign divided farmers' movements and reduced the political capability of NGOs supporting farmers' rights and environmental protection. Contested science drives different segments of the active political body in opposite directions. Interests in transgenics must be processed through a complex cognitive screen in which much that needs to be known is not known: the great ideational divide follows neither class nor North/South lines (Herring 2007). Intellectual property rights are reified as a force against the farmer – as if patents were somehow self-enforcing. Opponents fear the same reifications of bioproperty that techno-optimists posit as preconditions for progress. "Monopoly" by seed firms dominates the argument against transgenics even as stealth strategies defy property claims. Farmer-bred "Robin Hood" *Bt* seeds may be spreading faster than officially sanctioned seeds from "monopolist" Monsanto, as they are cheaper and often give better results (Roy, Herring, and Geisler 2007). Farmers in Brazil and India pursued the benefits of transgenic crops for reasons not credible, or even comprehensible, to their nominal allies in politics and NGOs.

Property rights in landscapes – in macro-nature – evince a long history of conflict between states and local communities, and are easier to claim in international treaties than in forests and fields (Herring 2002). Stealth seeds underscore the dependence of property rights in micro-nature on close monitoring mediated by high technology. In both Brazil and India, in different crops, technology developed by Monsanto was appropriated, redeployed, and developed by small-scale entrepreneurs and farmers themselves, unmindful of TRIPS negotiators or NGO petitions. In India, Delhi could not know that the cotton plants that survived the bollworm infestation of 2001 were transgenic and illegal prior to genetic testing.[39] Commercial interests diligently provided the knowledge state actors lacked; "seeing like a state" may be more astigmatic than seeing like a firm. Navbharat's successful *Bt* cotton violated no property rights, but clearly expressed the Monsanto CrylAc gene. In Brazil's transgenic soy, it is clear where the seeds originated, but it is less clear that Monsanto has a valid property claim after generations of field-level modification of varieties by farmers and seed firms operating independently of Monsanto. Bioproperty claims, like property claims generally, are relational; they mean what actors can make them mean. Ironically, the popular and successful Navbharat 151 seeds were banned – and their germplasm went underground – not because of Monsanto's property claim but because they had evaded biosafety procedures. In effect, as angry farmers and seed producers in Gujarat protested, to the extent Mahyco-Monsanto had privileged market position and extracted rents from farmers, it was a function not of market power but of restriction of alternatives by biosafety regulators.

Whatever else the genomics revolution may bring, it will certainly bring higher levels of surveillance. Surveillance has real costs, and certainly developmental opportunity costs; good policy rests on analysis that is currently absent. The analytics are missing partly because of inevitably incomplete science at the knowledge frontier: there are known unknowns and unknown unknowns. Because the transgenic genie is out of the bottle and imposes itself willy-nilly on society, the cost and consequences of biosafety regimes need attention.

The Goldilocks Paradox: Getting Regulation "Just Right"

To be normatively acceptable, development policy must argue for priorities at the margin: what are the opportunity costs of an extra dollar spent on biosafety regimes? How do benefits compare with alternative uses of that same dollar? The answer depends on how one conceptualizes benefits, how one couches alternatives, the normative position one takes on uncertainty and risk, and the projections one makes from an inevitably incomplete science. Figure 4.1 sketches the analytics of this problem. The first question is whether additional risk is introduced into plant breeding by recombinant-DNA (rDNA) technology compared with conventional breeding techniques. The answers in the literature vary from zero to a lot. Gene flow is pervasive in plants; all transfers of genetic material through any form of breeding entails some risk, however remote. There may be little if any *incremental* risk in genetic engineering. Moreover, invasive species allow whole genomes to flow through ecologies; these risks may dwarf the risk of single-gene flow, especially since not all transgene flow will improve the fitness of receptive plants: most crosses and mutations are less, not more, fit. Yet, control of invasive species seldom receives the level of rhetorical or financial support given to regulation of transgenics, despite enormous costs (Herring 2007).

If there is no incremental risk in rDNA techniques of plant breeding compared with alternatives, all expenditures – human capital, administrative time, institution building, money – will be wasted, whether the biosafety

Figure 4.1
Biosafety regime scenarios and social costs

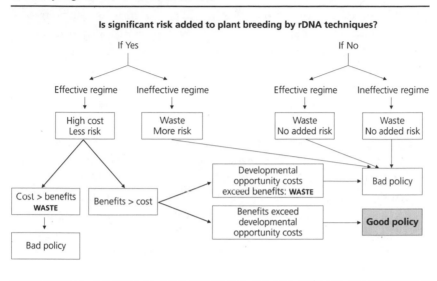

regime is effective or ineffective. The analogy might be to buying insurance against spells cast by enemies. Alternatively, let us assume that there *is* additional risk – the dominant position. It then matters whether the biosafety regime is effective or ineffective. If institutions are ineffective, incurred costs will go for naught: transgenes will spread and the outcome is again waste. An analogy might be the very expensive "war on drugs" periodically proclaimed by American administrations. Third, consider the situation where there *is* additional risk, and the biosafety regime is effective. Here analysis faces a new challenge: should the additional costs be incurred? Who should pay? If costs exceed benefits, and the biosafety regime is nevertheless established and nurtured, there is again waste. If benefits exceed costs, a second level of analysis becomes necessary: if the opportunity costs of the effective regime exceed the costs of the regime, we are again in the world of waste. It is *only* if the benefits of a biosafety regime clearly exceed the value of alternative uses of resources that there would be good policy.

It is hard to avoid the conclusion from the outcomes in Figure 4.1 that justifiable developmental policy promoting genetic engineering would be restricted to extremely important objectives, or restricted to deployments with very low risk. For low-income countries, opportunity costs are large: microbiologists, lab space, research protocols, administrative time, and money. Establishing a matrix of priorities would demand far more science than is currently available. By this logic, pro–vitamin A "golden rice" is in a different category from the herbicide-resistant grass for golf courses – recently shown to experience gene flow over very long distances in the United States.[40] A developmental calculus between fewer blind children and fewer weeds on golf courses might seem a simple one, but defenders of transgenic grass argue that weeds will be suppressed on golf courses whether we like it or not, and typically with much more dangerous chemicals than glyphosate, thereby justifying RR grass.

Not only are benefits socially disputed, but it is even difficult in principle to calculate risks (Thies and Devare 2007). *Bt* gene flow from cotton is unlikely to pose great risk. Test trials on *Bt* cotton found that there was no incremental risk: *Bt* plants showed the same characteristics as existing cottons in food and feed, except for the Cry1Ac protein, which disappears in pressed oils and seems to be harmful only to Lepidoptera. There was no evidence of damage to field ecology. Are these tests adequate? Could data from comparable regions of China or the United States have been used to ascertain risk, bypassing field trials in India? Or are even more fine-grained tests necessary, as agro-ecologies vary continuously rather than clumping into blunt categories?

There is also the risk that resistance to the endotoxin will develop in Lepidoptera – a threat to a public good deployed for decades by organic farmers who use *Bt* foliar sprays to control insects. Refugia are proposed for slowing the development of insect resistance to *Bt* toxin. These refugia are costly for small farmers and are often evaded. Is this a dangerous outcome? Every agro-ecology is slightly different. Mahyco presented to regulators the argument that there may actually be no problem with resistance development at all, as the bollworms will always have natural refugia in a great variety of noncotton plants; there are about 157 hosts for bollworms in India. This argument, while biologically plausible, failed because "uproar about the terminator" made authorities cautious.[41]

Capital in biotechnology complains that there is too much regulation – even though regulations favour deep pockets over small firms.[42] Environmentalists typically argue that there is too little. How would one recognize the Goldilocks standard of "just right": not too much, not too little?

Consider the cost of vetting *Bt* cotton in India. Although critics complained that biosafety testing was done in private and supervised by Mahyco-Monsanto, there were twelve agricultural universities involved in the studies, eight Central Research Institutes, the Indian Council for

Agricultural Research, the Indian Council for Medical Research, four Central Ministries and departments of agriculture in six states.[43]

From discussions with Monsanto beginning in 1993, an application by Mahyco to import seed was made on March 10, 1994; a permit was received on October 3, 1995. Mahyco imported Bollgard seed in 1996. Imported germplasm was backcrossed into Indian varieties. During this period, there were greenhouse and laboratory trials; these and later trials were mandated by the Review Committee on Genetic Manipulation (RCGM) of the Department of Biotechnology. There was only one limited field trial at one location. In 1997–98, a gene-flow study was conducted in limited field trials at five locations. In the same period, toxicological studies were conducted at the Industrial Toxicology Research Centre (ITRC) in Lucknow. In 1998–99, field and experimental trials were allowed by the RCGM at forty locations to test allergenicity and toxicity. Regulation in India is based on incremental risk – that is, risk added by the transgenic variety that is not present in the isogenic variety of the same plant.[44] For this type of trial, careful controls matching varieties are critical. These were the trials targeted by Operation Cremate Monsanto and from which the dissident farmer Mahalingappa conducted his own test of the terminator hypothesis. All safety data were submitted to the RCGM; the uprooting and burning protests induced authorities to request more trials. In 1999, multilocation trials replicated contained-research trials, at eleven locations. Procedures at this point reflected social turmoil; more caution was applied. It was only in 2000 that the RCGM recommended to the Genetic Engineering Approval Committee of the Ministry of Environment large-scale contained-field trials, which entailed seed production for testing. At this point, Navbharat transgenic seeds had been in fields for at least one year and were spreading to new acreage, but no regulatory authorities knew it.

Turmoil surrounding court cases and popular resistance caused the GEAC to ask for more field trials under auspices of the Indian Council for Agricultural Research (ICAR); these trials were planted late and thus were tainted in the eyes of NGO opponents, as peak bollworm infestation was avoided. Consequently, in 2001, a second year of field trials was ordered by the GEAC under the supervision of the ICAR. Before these results could be vetted, the "Robin Hood seeds" appeared in Gujarat, noticeable only because a bollworm rampage had wiped out other varieties. In 2002, field trial results were submitted to the GEAC, which subsequently, in March 2002, approved three *Bt* hybrids for commercialization. This decision was widely and correctly interpreted as a fait accompli.

Is this level of social expenditure justified by results? What was lost in these procedures? Excessive pesticides entered soils, water, and skins, with an unknown cost in poisonings and loss of fauna. The opportunity costs for Mahyco may have been high: presumably this time and effort could have gone into finding better hybrids than the original three, which were not the best available. Mahyco holds its cash costs to be approximately $8 million for development and testing of three transgenic varieties. The same transgene had been present in cotton in other countries for years, with no known ill effects. Navbharat bypassed biosafety procedures and created a *Bt* hybrid with early flowering characteristics well suited to Gujarati conditions; if Bollgard varieties are safe, so too is Navbharat 151 and all its variants/descendants – but the former are legal whereas the latter are not.

One future consequence of the stealth-seed phenomenon is predictable: real terminator technology. Social institutions will not deal with stealth seeds very well, as they do not deal with any high-value product that is movable across permeable state space: software, pornography, information, drugs, arms. Ajit Singh, the minister of agriculture at the time of the *Bt* discovery, said: "See that we cannot even regulate adulterated food; how will we regulate seeds?"[45] Gene use–restriction technologies (GURT) will become the only feasible way of restricting gene flow where there is risk, and the only way to enforce property claims under many conditions. The biological remedy is arguably more effective than the institutional remedy, if it can be made to work in the field.[46] Yet this conclusion illustrates again the power of discourse to shape and filter interest. Terminator technology was sidelined because of biopolitics. Is terminator technology acceptable as social policy? The clear answer has been no.

Disputes about material interests are demonstrably subject to conciliation, bargaining, and compromise – familiar terrain for the political economy of interests. Deals can be struck. Disputes about the nature of the natural, and consequent risks of the unnatural, take on a different politics, dependent on expertise that is asymmetrically distributed both locally – on the ground, within movements – and globally. Biopolitics is less susceptible to ordinary bargaining solutions, but it is an inescapable politics of the genomics revolution.

Acknowledgments
This chapter owes much to discussions with numerous political activists, seed producers, and farmers in India, as well as helpful interlocutors, of whom only a few can

be mentioned here: Sivramiah Shantharam, Suman Sahai, Kameswara Rao, D.B. Desai, Vikas Chandak, Anil Gupta, Michael Lipton, Maria Jose Sampaio, Andre Goncalves, and numerous colleagues at Cornell University.

Reprinted by permission of the publisher (Taylor & Francis Ltd, http://www.tandf online.com), Herring, Ronald J. 2007. "Stealth Seeds: Bioproperty, Biosafety, Bio-politcs," *Journal of Development Studies* 43 (1): 130–57.

Notes

1 See generally Herring 2003a on the historical uniqueness of agrarian struggles in Kerala. The Monsanto-terminator story circulated even in India's premier journal of the social sciences, the *Economic and Political Weekly:* Shiva, Emani, and Jafri 1999; Sahai 1999.

2 On the conjunction between globalization and terminator technology, Suman Sahai, leader of Gene Campaign, wrote (1999, 84): "Trials of a genetically altered cotton variety (*Bt* cotton) conducted by the American company Monsanto have provided the trigger because Monsanto also happens to own the terminator technology." See also Bharathan 2000; Parmar and Visvanathan 2003. RAFI International 2000 seems to be the origin of the phraseology in its communiqué "Terminator 2 Years Later: Suicide Seeds on the Fast Track." See generally Gold 2003 on terminator discourse.

3 On debt as a cause of suicides, see Centre for Environmental Studies Warangal 1998; Department of Agriculture and Cooperation 1998. The Government of Karnataka (2002) concluded that debt and agrarian crisis were implicated, but many suicides were tragically personal in nature.

4 How much leaders knew about the truth of the discourse is uncertain. Vandana Shiva, A. Emani, and A.H. Jafri wrote in 1999 of the seeds: "They are in an ecological sense terminator, which terminates biodiversity." Shiva and colleagues (2000, 98) wrote: "Freedom from the second cotton colonisation needs to be based on liber-ation through the seed ... The freedom of the seeds [is] simultaneously a resistance against monopolies ... like Monsanto ... The seeds of suicide need to be replaced by the seeds of prosperity."

5 See also Greenpeace-India 2001; Jayaraman 2001; Mathew 2001; Sahai 2002; Shaik 2001; Mehta 2005 (60–79, 130–36); Visvanathan and Parmar 2002; Scoones 2003; Omvedt 2005. These sources are complemented by the author's discussion with Mr. Raju Barwale, managing director of Mahyco, May 28, 2004, and Dr. D.B. Desai of Navbharat and *Bt*-cotton farmers, Gujarat, June 2005.

6 See generally Government of India 2002a and 2002b. This information was com-plemented by the author's interviews in New Delhi (June 2002) and Ahmedabad (2005).

7 See generally Gene Campaign 2005, pt. 2, 72 et passim. This information was com-plemented by the author's interviews in Delhi (2002) and Ahmedabad (2005).

8 Minister Patil told the State Legislative Council that this outcome was assured by the Agriculture Secretary in Delhi, a result of successful lobbying by the state govern-ment. The question was raised by N.P. Hirani, a legislator who is also administrator of the State Cotton-Growers Association (*Indian Express,* December 11, 2001).

9 See also *Indian Express* (Mumbai/Nagpur edition), November 13, 2001, (Delhi edition), March 25, 2002; *Financial Express,* April 3, 2002; Gene Campaign 2005, pt. 2, 101, 109 et passim.

10 See Yamaguchi 2004 (ch. 4), who found that farmers with relatives in Andhra Pradesh were more successful in obtaining Navbharat 151 seeds after they were discovered and banned. The text's account was confirmed by conversations between the author and seed producers in Gujarat (June 2005).

11 The English is sometimes used in rural Gujarat to designate what farmers also call *lodhavela biyaran* ("ginned seeds") to designate nonbranded F2 *Bt* cotton seeds.

12 I have heard farmers insist that F2s outperform official seeds. Biologically, this seems unlikely, but it is possible that variable agronomic conditions favour variance in phenotypic characteristics other than those for which the hybrid was chosen. See Roy, Herring, and Geisler 2007, on farmer logics of seed choice.

13 Personal communication, October 16, 2004; see also SeedQuest 2003a; *Indian Express,* May 28, 2003; Jayaraman 2004b; *The Hindu,* November 4, 2004.

14 See Government of India 2002b; Gene Campaign 2005, pt. 2, 102–3; author's interview with Ajit Singh, former minister of agriculture, New Delhi, June 22, 2005.

15 For positive agronomic results in Gujarat, see Roy, Herring, and Geisler 2007; Morse, Bennett, and Ismael 2005. Mahyco-Monsanto's own data for the first year of legal production of their MECH varieties on 55,000 farms show an all-India average reduction in pesticide use per acre of 65–70 percent, worth 920 rupees to the average farmer, an increase in yields of 3.22 quintals per acre, and an increase in income of 7,520 rupees per acre (unpublished data). The smallest benefits were reported in Andhra Pradesh (5,930 rupees), where much of the "failure of *Bt* cotton" literature originates. O.M. Bambawale and colleagues (2004), in an independent assessment of MECH-162, find positive results. M. Qaim and D. Zilberman (2003) created a stir for projecting atypically positive *Bt* cotton results from India.

16 See *The Hindu,* December 14, 1998. Compare Rediff.com 2001 and Jha 2001. Even the serious journal *Frontline* carries similar stories: A. Krishnakumar (2004) confidently analyzed "the disaster wrought by *Bt* cotton." For useful commentary, see Rao 2004a. Claims of failure began with field trials: Shiva, Emani, and Jafri 1999. Reports of *Bt* failures on a few farms even merited parliamentary discussion. Responding to the insistence of an MP from Gujarat, a central team of scientists was sent to investigate MECH-162; the team found normal agronomic variance (Mistry 2003).

17 The cultivar MECH-184, for example, needs early watering to develop; otherwise, it wilts. O.M. Bambawale and colleagues (2004) report unpublished data showing superior performance of even this demanding variety, a finding consistent with Roy, Herring and Geisler 2007. Although G.D. Stone (2004) argues that biotechnology is especially "de-skilling," conventional hybrids often require new knowledge and techniques.

18 Interview with Mahalingappa, June 7, 2004. See Madsen 2001, 3733–42; Birasal 1998; Visvanathan and Parmar 2002.

19 *The Hindu,* December 14, 1998. Similar accounts appear in *Deccan Herald* 1998 (Bangalore edition).

20 Interviews with MMBL, GEAC, and cotton-seed producers.

21 Meera Nanda (2003) argues that anti-Enlightenment epistemological relativism, manifest in "alternative" sciences, is instrumentally deployed by the *Hindutva* right as well as by public intellectuals claiming the mantle of the left.

22 Press release of June 20, 2003, New Delhi. Signatories of the petition range ideologically from the All-India Bhartiya Kisan Sangh, a Bharatiya Janata Party (BJP) affiliate, to the All India Kisan Sabha, a Communist Party of India affiliate.

23 Mehta 2005, 60–76; interviews with Dr. D.B. Desai and the breeder of Navbharat 151, Dr. N.P. Mehta, in Ahmedabad, June 2005.

24 Conversations of author with Percy Schmeiser, January 27–28, 2004. For a brief history, see Federal Court of Appeal decision in *Monsanto Canada Inc. v. Schmeiser,* 2002 FCA 309, appeal ref'd 2004 SCC 34.

25 The text also relies on various issues of *Actualización: Un boletín de noticias & análisis mensuales de la Campaña Por Un Brasil Libre de transgênicos* (hereafter *Actualización*).

26 Ironically, the state of Rio Grande do Sul banned cultivation of all transgenic crops effective January 1, 2000 (Niiler 1999, 848). On early decisions, see Hathaway 2002; Fontes 2003.

27 Sources for this section include *Actualización;* Niiler 1999; Maria Jose Sampaio, personal communications, 2003, 2004.

28 Estimates of planted area, here and later, are based on extrapolation from area harvested backward to seeds sold; nontransgenic seeds sold were sufficient to plant only about 40 percent of the state's area (Benson 2001). Saved seeds, transgenic and otherwise, are not counted by this technique. See also Nuffield Council on Bioethics 2004, 3.57–3.61.

29 "En resumo, no había ningún otro objetivo para esta decisión de ahora sino reforzar la estrategia de situación de facto de los pro-OGMs." *Actualización,* September 25, 2003, Issue 3.

30 INASE opposed "brown bag" seeds that carry "no guarantee of quality and identity ... and royalties are not paid." The state's interests are engaged because no taxes are collected on underground sales (Elliott 1999).

31 Ironically, the seeds were produced by local farmers, not Monsanto (Poddar 2004, 32; Organic Consumers Association 2001).

32 Numerical estimates of farmers who have registered to plant transgenic soy vary so much that reporting is premature; there is both compliance and evasion (Ewing 2003).

33 The mayor of Chapada, Rio Grande do Sul, Carlos Alzenir Catto, was quoted as saying that "98 per cent" of the farmers in his area, including small farmers, were registered to plant transgenic soy: "Here, people don't have any fear of signing up because they know they are planting a product that is not bad for health or the environment, because they can spray less agrochemicals on the crops" (Ewing 2003).

34 Lewis 2004; Reuters 2004a.

35 Leila Macedo Oda, president of the NGO ANBio (National Biosafety Association), draws a parallel with the "market reserve" in computing in the 1980s. By forbidding import of computers, the Brazilian government hoped to favour the local computer industry, but lack of competition only made Brazil lag behind other countries.

Computers were then smuggled, much as farmers now smuggle seeds (Bonalume Neto 2003, 1257–58).

36 The Agriculture Department of Paraná cordoned fields near the town of Pato Branco when crops tested positive for transgenic soy. The crop was ruled illegal because the owner had not signed an agreement to observe new rules for transgenic cultivation. Only 464 farmers had signed the agreement – itself contradicted by a state law, later overruled, banning transgenics. Many farmers refused to sign because the agreement failed to reflect modifications made by Congress to the biosafety bill. The agreement requires that farmers pay royalties on illicitly produced seeds; the new bill rejects this liability (Ewing 2003; American Soybean Association 2004; Clendenning 2004).

37 Salazar et al. 2000; Pray and Schimmelpfenning 2001; Pray and Naseem 2007; Jayaraman 2004b. Luisa Massarani (2005) reported an accusation by the state deputy Frei Sérgio Antônio Görgen of transgenic corn being sold in Rio Grande do Sul. Greenpeace announced in mid-April 2005 findings of a German laboratory of *Bt* genes in rice seed in markets near Wuhan (New Agriculturist On-line 2005).

38 The parallel to James Scott's *Seeing Like a State* (1998) is clear; Scott's high modernist state needs legibility of its terrain, but finds it hard to attain. The Panopticon is posited, but illusive.

39 There are now much simpler *Bt* detection kits developed in India, costing about 50 rupees (US$1) per kit.

40 Andrew Pollack (2004) reports on a study forthcoming in *Proceedings of the National Academy of Sciences* of transgenic bentgrass created by Monsanto and Scott for golf courses for weed management. Gene flow extended as far as twelve miles, which was unexpected. The US Forest Service is quoted as saying that the grass "has the potential to adversely impact all 175 national forests and grasslands."

41 Interview with Mr. Raju Barwale, May 28, 2004. MMBL was required to provide seed for the refugia – in a packet of 450 grams of *Bt* seed, 120 grams of non-*Bt* isogenic seed is supplied by the firm for refugia planting, along with instructions. The firm is expected to do resistance studies. Cited sources for this section are supplemented by interviews in Delhi, July 2001, June 2002, June 2005. See also Department of Biotechnology 1996.

42 For an argument that approval is excessively slow for biotech in India, see Confederation of Indian Industry 2001.

43 Sources for this section include conversations with Mahyco officials, officers of the Department of Biotechnology, NGO activists, and Department of Biotechnology 2002; Scoones 2003; Bagla 2004.

44 Transgenics are governed by the *Environment (Protection) Act of 1986*, Rules of 1989, and specifically Notification No. G.S.R.1037(E) dated December 5, 1989.

45 Interview, New Delhi, June 21, 2005.

46 By analogy, a forestry official in Bangladesh once told me: "Tigers are the best forest guards – utterly incorruptible."

References

Legislation
Environment (Protection) Act, 1986, No. 29 of 1986 (May 23, 1986) [India].

Law No. 8.078 of September 11, 1990 – *Código de Proteção e Defesa do Consumido* [Consumer Defence Code] [Brazil].

Law No. 9.279 of May 14, 1996 – *Código de Propriedade Industrial* [Industrial Property Code] [Brazil].

Rules for the Manufacture/Use/Import/Export and Storage of Hazardous Microorganisms, Genetically Engineered Organisms or Cells, Notification No. G.S.R. 1037(E), December 5, 1989 [Rules of 1989] [India].

Seeds Act, 1966, No. 54 of 1966 (December 29, 1966) [India].

Case

Monsanto Canada Inc. v. Schmeiser, 2002 FCA 309, appeal ref'd 2004 SCC 34.

Periodicals

Actualização: Um boletim de noticias e análises mensais de Campanha por um Brasil Livre de Transgênicos

Financial Express

Indian Express

The Hindu

Secondary Sources

American Soybean Association. 2004. "Weekly Update." *Archives,* 19 January.

Associação Nacional de Biossegurança. 2001. "Report on the Commercial Approval of Transgenic Soy in Brazil." Associaçiao Nacional de Biossegurança, http://www. anbio.org.br.

Bagla, P. 2004. "Report Says India Needs Stronger, Independent Regulatory Body." *Science* 304 (5677): 1579. http://dx.doi.org/10.1126/science.304.5677.1579a.

Bambawale, O.M., A. Singh, O.P. Sharma, et al. 2004. "Performance of *Bt* Cotton (MECH-162) under Integrated Pest Management in Farmers' Participatory Field Trial in Nanded District, Central India." *Current Science* 86 (12): 1628–33.

Benson, Todd. 2001. "GM Soy Planting in Brazil Said to Be Spreading North." *Dow Jones,* December 18.

Bharathan, G. 2000. "*Bt*-Cotton in India: Anatomy of a Controversy." *Current Science* 79 (8): 1067–75.

Birasal, N.R. 1998. "Haveri Farmers Will Resist KRRS Destruction Trial." *Times of India,* December 5.

Bonalume Neto, R. 2003. "GM Confusion in Brazil." *Nature Biotechnology* 21 (11): 1257–58. http://dx.doi.org/10.1038/nbt901.

Buckly, S. 2000. "Brazil Battles over Genetically Modified Soybeans." *Washington Post,* June 27.

Centre for Environmental Studies Warangal. 1998. *Citizens' Report: Gathering Agrarian Crisis – Farmers' Suicides in Warangal District (AP) India.* Kishanpura: Centre for Environmental Studies Warangal.

Clendenning, A. 2004. "Illegal Soy Seeds Enrich Brazil Farms." *Washington Times,* January 6.

Confederation of Indian Industry. 2001. "Biotechnology on the Fast Track." Unpublished conference proceedings, New Delhi.

Conroy, John. 2003. "A Time to Reap." *Spiked Online,* November 13. http://www. genet-info.org.

Deccan Herald. 1998. "Operation Cremate Monsanto: Raitha Sangha to Burn Bollgard Cotton in Bellary." *Deccan Herald* (Bangalore), December 2.

Department of Agriculture and Cooperation, Ministry of Agriculture, Government of India. 1998. *Report of the Study Group on Distress Caused by Indebtedness of Farmers in Andhra Pradesh.* New Delhi: Government of India.

Department of Biotechnology, Ministry of Science and Technology, Government of India. 2002. *Annual Report 2001–2002.* New Delhi: Government of India.

–. 1996. *Revised Guidelines for Research in Transgenic Plants and Guidelines for Toxicity and Allergenicity Evaluation of Transgenic Seeds, Plants and Plant Parts.* New Delhi: Government of India.

DKSS (Deshiya Karshaka Samrakshana Samithi). 2003. *National Farmers' Protection Committee.* Palakkad, India: Plan and Budget.

Dow Jones. 1998. "Monsanto Denies Testing Seed with High-Yield Gene in India." *Agnet,* November 20.

Elliott, Robert S. 1999. "Illegal GM-Crop Trading in Brazil and Argentina." *GENET,* September 28 and 30. http://www.gene.ch/genet/1999/Oct/msg00008.html.

Ewing, Reese. 2003. "Brazil Farmers Declare Plans for Genetically Modified Soybeans." *Reuters,* November19.

Fontes, E.M.G. 2003. "Legal and Regulatory Concerns about Transgenic Plants in Brazil." *Journal of Invertebrate Pathology* 83 (2): 100–3. http://dx.doi.org/10.1016/S0022-2011(03)00060-0.

Gene Campaign. 2005. *The Story of Bt Cotton in India.* CD-ROM. New Delhi: Gene Campaign.

Ghosh, P.K. 2001. "Genetically Engineered Crops in India with Special Reference to *Bt* Cotton." *IPM Mitr* 2: 8–21.

Gold, A.G. 2003. "Vanishing: Seeds' Cyclicality." *Journal of Material Culture* 8 (3): 255–72. http://dx.doi.org/10.1177/1359183503008300 2.

Government of India. 2002a. "Cultivation of Bt Cotton Using Navbharat Seeds." Unstarred Question No. 205 to Be Answered on 01.03.2002 by Minister for Environment and Forests, Shri T.R. Baalu. Rajya Sabha.

–. 2002b. "Restriction of Production and Sale of *Bt* Cotton." Unstarred Question No. 206 to Be Answered on 01.03.2002 by Minister for Environment and Forests, Shri T.R. Baalu. Rajya Sabha.

Government of Karnataka. 2002. *Farmers' Suicides in Karnataka: Scientific Analysis Report of the Expert Committee for Study on Farmers' Suicides.* Bangalore: Office of the Expert Committee for Study on Farmers' Suicides.

Greenpeace-India. 2001. "Bt Cotton Saga Takes a Ludicrous Turn." *Greenpeace Newsletter* (New Delhi) 11 (December).

Gupta, A.K., and V. Chandak. 2005. "Agricultural Biotechnology in India: Ethics, Business and Politics." *International Journal of Biotechnology* 7 (1/2/3): 212–27. http://dx.doi.org/10.1504/IJBT.2005.006455.

Hathaway, D. 2002. "Legal Controls over GMOs in Brazil." *Actualización: Un boletín de noticias & análisis mensuales de la Campanha Por Un Brasil Libre de transgênicos* [Updates: A Bulletin of News and Analysis of the Campaign for a Brazil Free of Transgenics], February 25.

Herring, R.J. 2000. "Authority and Scale in Political Ecology: Some Cautions on Localism." In *Biological Diversity: Balancing Interests through Adaptive Collaborative Management,* ed. L. Buck, C.C. Geisler, J. Schelhas, et al., 187–205. Boca Raton, FL: CRC Press.

–. 2002. "State Property Rights in Nature (with Special Reference to India)." In *Land, Property and the Environment,* ed. J.F. Richards. Oakland, CA: Institute for Contemporary Studies.

–. 2003a. "The Political Impossibility Theorem of Agrarian Reform: Path Dependence and Terms of Inclusion." In *Changing Paths: The New Politics of Inclusion,* ed. M. Moore and P.P. Houtzager, 58–87. Ann Arbor: University of Michigan Press.

–. 2003b. Data as a social product. In *Q-Squared: Combining Qualitative and Quantitative Methods in Poverty Appraisal,* ed. R. Kanbur. New Delhi: Permanent Black.

–. 2005. "Miracle Seeds, Suicide Seeds and the Poor: GMOs, NGOs, Farmers and the State." In *Social Movements in India: Poverty, Power, and Politics,* ed. R. Ray and M.F. Katzenstein, 203–32. Lanham, MD/New Delhi: Rowman and Littlefield/Oxford University Press.

–. 2007. "The Genomics Revolution and Development Studies: Science, Poverty and Politics." *Journal of Development Studies* 43 (1): 1–30. http://dx.doi.org/10.1080/00220380601055502.

Hochstetler, K. 2004. *Civil Society in Lula's Brazil.* Working Paper No. CBS-57–04, Centre for Brazilian Studies, University of Oxford. Oxford: Latin American Centre, Oxford University. http://www.lac.ox.ac.uk/sites/sias/files/documents/Kathryn%2520Hochstedler%252057.pdf.

James, C. 2002. *Global Review of Transgenic Crops: 2001 Feature: Bt Cotton.* ISAAA Briefs No. 26. Ithaca, NY: International Service for the Acquisition of Agri-Biotech Applications.

Jayaraman, K.S. 2001. "Illegal Bt Cotton in India Haunts Regulators." *Nature Biotechnology* 19 (12): 1090. http://dx.doi.org/10.1038/nbt1201-1090.

–. 2004a. "Illegal Seeds Overtake India's Cotton Fields." *Nature Biotechnology* 22 (11): 1333–34. http://dx.doi.org/10.1038/nbt1104-1333.

–. 2004b. "India Produces Homegrown GM Cotton." *Nature Biotechnology* 22 (3): 255–56. http://dx.doi.org/10.1038/nbt0304-255.

Jha, S. 2001. "Seeds of Death, GMO Cotton." *India,* May 18.

Joshi, S. 2001. "Unquiet on the Western Front." *The Hindu Business Line,* December 19.

Krishnakumar, A. 2004. "Bt Cotton, Again." *Frontline* 21 (10): 8–21. AgBioWorld, http://www.agbioworld.org/newsletter_wm/index.php?caseid=archive&newsid=2117.

Lewis, S. 2004. "Brazil's Biosafety Law Stalls in the Senate." *Food Chemical News* (USA), June 14.

Liptak, A. 2003. "Saving Seeds Subjects Farmers to Suits over Patent." *New York Times,* November 2, 1.18.

Madsen, S.T. 2001. "The View from Vevey." *Economic and Political Weekly* 36 (39): 3733–42.

Massarani, Luisa. 2005. "Illegal GM Corn Found in Brazil." SciDev.net, December 2.

Mathew, V. 2001. "India's GM Cotton Story Gets Bigger – 'Uproot & Destroy' Begins on Gujarat Farms." *The Hindu Business Line*, October 20.

Mehta, N.P. 2005. "Genesis of Cotton Hybrid Navbharat-151 and Bt Cottons and Their Prospects." In *Bt Cotton: A Painful Episode*, ed. S. Mehta. Vadodara, Gujarat: Gujarat Kapas Utpadak Hitrakshak Sangh.

Mehta, S., ed. 2005. *Bt Cotton: A Painful Episode.* Vadodara, Gujarat: Gujarat Kapas Utpadak Hitrakshak Sangh.

Mistry, M. 2003. Madhusudan Mistry, MP, Lok Sabha, letter to Dr. Murli Manohar Joshi, Ministry of Science and Technology, October 18, 2003; answered in Parliament December 8, 2003, by Mom Raj Singh, Under Secretary, NO.RS.9/6/2003-S&T.

Mistry, S. 1998. "Terminator Gene a Figment of Imagination: Monsanto Chief." *Indian Express*, December 4.

Morse, S., R. Bennett, and Y. Ismael. 2005. "Comparing the Performance of Official and Unofficial Genetically Modified Cotton in India." *AgBioForum: The Journal of Agrobiotechnology Management and Economics* 8 (1): 1–6. http://agbioforum.org/v8n1/v8n1a01-morse.pdf.

Naik, G., M. Qaim, A. Subramanian, et al. 2005. "Bt Cotton Controversy: Some Paradoxes Explained." *Economic and Political Weekly* 40 (15): 1514–47.

Nanda, M. 2003. *Prophets Facing Backward: Postmodern Critiques of Science and Hindu Nationalism in India.* New Brunswick, NJ: Rutgers University Press.

New Agriculturist On-line. 2005. "News Brief." May 1. New Agriculturist On-line, http://www.new-agri.co.uk/05-3/newsbr.html#nb7.

Niiler, Eric. 1999. "Monsanto Remains a Magnet for GM Opposition." *Nature Biotechnology* 17 (9): 848. http://dx.doi.org/10.1038/12831.

Nuffield Council on Bioethics. 2004. *The Use of Genetically Modified Crops in Developing Countries.* London: Nuffield Council on Bioethics.

Omvedt, G. 2005. "Farmers' Movements and the Debate on Poverty and Economic Reforms in India." In *Social Movements in India: Poverty, Power, and Politics*, ed. R. Ray and M.F. Katzenstein, 179–202. Lanham, MD/New Delhi: Rowman and Littlefield/Oxford University Press.

Organic Consumers Association. 2001. "Brazilian Farmers Seize Monsanto Facilities in Anti-GE Protest." January 26. Organic Consumers Association, http://www.organicconsumers.org/monsanto/brazilprotest.cfm.

Osava, M. 2002. "Transgenic Soy Rampant, Despite Ban." *Inter Press Services*, April 5. http://www.ips.org.

–. 2004. "Transgenic Soy Found Guilty by People's Court." *Inter Press Services*, March 11. http://www.ips.org.

Paarlberg, R.L. 2001. *The Politics of Precaution: Genetically Modified Crops in Developing Countries.* Baltimore: Johns Hopkins University Press, in cooperation with the International Food Policy Research Institute.

Parmar, C., and S. Visvanathan. 2003. "Hybrid, Hyphen, History, Hysteria: The Making of the Bt Cotton Controversy." Institute of Development Studies seminar on Agriculture Biotechnology and the Developing World, Sussex, UK, October 1–3. http://www.ids.ac.uk/files/SVCPCotton.pdf.

Poddar, U. 2004. "Controversy over Genetically Modified Organisms: A Comparative Study of Brazil and India." Master of Professional Studies (MPS) thesis, Cornell University.

Pollack, A. 2004. "Genes from Engineered Grass Spread for Miles, Study Finds." *New York Times,* September 21.

Pray, C., and D. Schimmelpfennig. 2001. "The Impact of Biotechnology on South African Agriculture: First Impressions." College of Agriculture seminar, University of Pretoria, South Africa.

Pray, C.E., J. Huang, R. Hu, et al. 2002. "Five Years of Bt Cotton in China – the Benefits Continue." *Plant Journal* 31 (4): 423–30. http://dx.doi.org/10.1046/j.1365-313X.2002.01401.x.

Pray, Carl, and Anwar Naseem. 2007. "Supplying Crop Biotechnology to the Poor: Opportunities and Constraints." *Journal of Development Studies* 43 (1): 192–217. http://dx.doi.org/10.1080/00220380601055676.

Qaim, M., and A. de Janvry. 2003. "Genetically Modified Crops, Corporate Pricing Strategies, and Farmers' Adoption: The Case of Bt Cotton in Argentina." *American Journal of Agricultural Economics* 85 (4): 814–28. http://dx.doi.org/10.1111/1467-8276.00490.

Qaim, M., and D. Zilberman. 2003. "Yield Effects of Genetically Modified Crops in Developing Countries." *Science* 299 (5608): 900–2. http://dx.doi.org/10.1126/science.1080609.

RAFI International. 2000. "Terminator 2 Years Later: Suicide Seeds on the Fast Track." RAFI communiqué of February-March and news release of February 20, Winnipeg, MB.

Rao, C.K. 2004a. "Frontline Mischief." *AgBioView* electronic newsletter, May 18. AgBioWorld, http://www.agbioworld.org/newsletter_wm/index.php?caseid=archive&newsid=2117.

–. 2004b. "One Swallow Does Not Make the Summer – Bt Cotton in India." Foundation for Biotechnology Awareness and Education (Bangalore), http://fbae.org/2009/FBAE/website/special-topics_one_swallow_does_not_make_the_summer.html.

Rediff.com. 2001. "Bt Cotton Will Kill Farmers, Financially and Literally." Interview with Devinder Sharma, December 12. Rediff.com, http://www.rediff.com/money/2001/dec/12inter.htm.

Research Foundation for Science, Technology and Ecology. 2003. "Beyond Bhopal and *Bt:* Taking on the Biotech Giants." Asian Social Forum seminar, Hyderabad, India, January 4.

Reuters. 2001. "Brazilian Farmers Seize Monsanto Facilities in Anti-GE Protest." Não Me Toque, Brazil, January 26. Organic Consumers Association, http://www.organicconsumers.org/monsanto/brazilprotest.cfm.

–. 2003. "Brazil Suspends Anti-GM Law." December 11.

–. 2004a. "Brazil Farmers Fight for GMOs." February 12.

–. 2004b. "Brazilian Scientists Develop New GMO Soybean." March 10.

Roy, D., R.J. Herring, and C.C. Geisler. 2007. "Naturalizing Transgenics: Official Seeds, Loose Seeds and Risk in the Decision Matrix of Gujarati Cotton Farmers." *Journal of Development Studies* 43 (1): 158–76. http://dx.doi.org/10.1080/00220380601055635.

Sahai, S. 1999. "What Is Bt and What Is Terminator?" *Economic and Political Weekly* 34 (3/4): 84–85.

–. 2002. "Bt Cotton: Confusion Prevails." *Economic and Political Weekly* 37 (21): 1973–74.

Salazar, S., C. Falconi, J. Komen, et al. 2000. *The Use of Proprietary Biotechnology Research Inputs at Selected Latin American NAROs.* Briefing Paper No. 44. The Hague: ISNAR.

Scoones, I. 2003. "Regulatory Manoeuvres: The Bt Cotton Controversy in India." Working Paper No. 197. Brighton, UK: Institute of Development Studies.

Scott, J.C. 1998. *Seeing Like a State.* New Haven, CT: Yale University Press.

SeedQuest. 2003a. "Brazil Introduces Bill to Regulate GM Crops." Rio de Janeiro, October 31. eFeedLink, http://www.efeedlink.com/contents/10-31-2003/51447784-6921-4138-b097-ca857f178f76-b081.html.

–. 2003b. "GM Is Cottage Industry; Hybrid Seeds Flood Gujarat Fields." Rajkot/Junagadh, news release, May 28.

Shaik, Sajid. 2001. "Farmers Decide to Defend Their Bt Gene Cotton Crops." *Times of India,* October 31.

Sharma, D. 2001. "The Introduction of Transgenic Cotton in India." *Biotechnology and Development Monitor* (44): 10–13.

Shiva, V. 1997. *Biopiracy: The Plunder of Nature and Knowledge.* Boston: South End Press.

–. 1999. "Ecological Balance in an Era of Globalization." In *Global Ethics and Environment,* ed. N. Low, 47–69. London: Routledge.

Shiva, V., A. Emani, and A.H. Jafri. 1999. "Globalization and Threat to Seed Security: Case of Transgenic Cotton Trials in India." *Economic and Political Weekly* 34 (10/11): 601–13.

Shiva, V., A.H. Jafri, A. Emani, et al. 2000. *Seeds of Suicide: The Ecological and Human Costs of Globalisation of Agriculture.* New Delhi: Research Foundation for Science, Technology and Ecology.

Stone, G.D. 2002. "Biotechnology and Suicide in India." *Anthropology News* 43 (5): 5. http://dx.doi.org/10.1111/an.2002.43.5.5.2.

–. 2004. "Biotechnology and the Political Ecology of Information in India." *Human Organization* 63 (2): 127–40. http://dx.doi.org/10.17730/humo.63.2.jgvu7rlfafk9jwf9.

Thies, J.E., and M. Devare. 2007. "An Ecological Assessment of Transgenic Crops." *Journal of Development Studies* 43 (1): 97–129. http://dx.doi.org/10.1080/00220380601055593.

Visvanathan, S., and C. Parmar. 2002. "A Biotechnology Story: Notes from India." *Economic and Political Weekly* 37 (27): 2714–24.

Yamaguchi, T. 2004. "Discourse Perspective on Agrifood Biotechnology Controversies: Bt Cotton in India." PhD dissertation, Michigan State University.

Glossary

Beej panchayat – "seed council" (literally, village council of five, to govern seeds).

Biopolitics – politics concerning biological entities, such as genetic materials.

Bioproperty – intellectual property in biological entities, such as patents on seeds, or intellectual property rights in germplasm in general.

F2 generation – the generation resulting from the cross of two (F1) hybrids.

Navbharat / Navbharat Seeds – Seed company located in Ahmedabad, Gujarat State, India, responsible for the illegal Navbharat 151 *Bt* cotton seeds that proved successful but were prohibited by the Government of India's Genetic Engineering Approval Committee in 2001.

Robin Hood seeds – name given to the underground, illegal transgenic cotton seeds by some observers, implying illicit appropriation of intellectual property rights in favour of poor farmers.

Shetkari Sanghatana – Association of Agriculturalists; name of the then-largest farmers' movement in India, based in the state of Maharashtra.

Stealth seeds – seeds circulating under the radar of both firms and government agencies in networks of farmers; sold without biosafety certification and without intellectual property claims.

PART 2

INTELLECTUAL PROPERTY MECHANISMS AND COUNTER-MOVEMENTS

5

Implementing the *International Treaty on Plant Genetic Resources for Food and Agriculture*

A Regulatory and Intellectual Property Outlook

CHIDI OGUAMANAM

The discovery of recombinant DNA (rDNA) in the 1970s, and the gradual uptake of insights thereof in the realm of agricultural genomics[1] and broader biotechnologies, resulted in greater private sector interest in agriculture. As a consequence, stronger intellectual property protection is now seen as a sine qua non for the profitability of innovation, investment in the agricultural sector, and sustainability of private sector interest in agricultural biotechnology (ag-biotech). In the decades following the discovery of rDNA, a combination of several factors (Kloppenburg 1988) that are outside the scope of this chapter, resulted in decreased public sector and increased private sector funding for agricultural research and development (Evenson, Santaniello, and Zilberman 2002). Some analysts agree that this new dynamic in agricultural R&D promotes the development of commercially viable agricultural products and innovations, especially monocultures (Shiva 2000). The focus on economically viable monocultures is accompanied by diminished interest in agricultural biodiversity. It also tends to undermine the potential to adapt ag-biotech innovations to socially beneficial outcomes, such as, through a focus on commercially insignificant but socially beneficial crops and other genetic resources. Under this new innovation framework, alternative agricultural epistemologies, including traditional agricultural practices of indigenous and local communities (ILCs), are relegated to the bottom of agricultural R&D priorities (Oguamanam 2007).

While commercial considerations drive R&D in agricultural genomics, genomics technologies have the potential to deliver an increased range of socially beneficial agricultural outputs, including new adaptations of traditional varieties, although such outputs may also be of low commercial value. Despite their limited commercial benefits, such varieties have the potential to alleviate global hunger and positively affect vulnerable members of the human family in remote ILCs and elsewhere. Aside from the fact that such outcomes are not research priorities, the extent to which innovations in ag-biotech could contribute to socially beneficial outcomes is also a product of the undergirding regulatory environment, which ought to include a combination of open/shared and IP/proprietary mechanisms (Marden and Godfrey 2012). In this chapter, I focus on the implementation experience of the *International Treaty on Plant Genetic Resources for Food and Agriculture* (ITPGRFA) of the Food and Agriculture Organization of the United Nations (FAO) and other increasingly complex associated regimes (Chiarolla 2008a). These include regimes dealing with conservation of biodiversity, and access and benefit sharing over genetic resources and the utilization of their components. In some ways, even though these regimes do not discount the role of IP, they help to moderate or balance its generally restrictive effects on access to the benefits of research and innovation in agricultural genomics. I am interested in mapping the contemporary outlook on these interlocking regimes in order to understand the degree to which they are capable of advancing socially beneficial outcomes for agricultural genomics, especially in the context of the rapidly changing agricultural R&D landscape, where public/private collaborations remain inevitable.

The ITPGRFA

The ITPGRFA has a long history, dating back to before the 1980s. Its previous incarnation was the nonbinding *International Undertaking on Plant Genetic Resources for Food and Agriculture* (IUPGRFA) in 1983. A primary goal of the ITPGRFA and its precursor was to secure access for developing countries, and their indigenous and local farming communities, to innovations arising from the use of plant genetic resources, especially those in the ex situ (i.e., outside their natural origins) global "gene banks" or "seed banks." At the same time, the ITPGRFA aims to ensure that stakeholders in agricultural R&D have access to these important genetic resources. The seeds in these banks were voluntarily contributed by or sourced from countries (most of which are also centres of global biodiversity) pursuant to the principle that plant genetic resources are the common heritage of all humanity.

The seeds were held in trust by the Consultative Group for International Agricultural Research (CGIAR) through its federating International Agricultural Research Centres (IARCs) under the supervision of the FAO. Over the years, however, through the application of IP rules and various technological strategies, coupled with the increased visibility of the private sector within the CGIAR system, the management of plant genetic resources in the so-called global seed banks created a crisis of confidence and a dichotomy between suppliers and users of those resources corresponding to developing and developed country blocs. As R&D in ag-biotech advanced, especially in the last quarter of the twentieth century, so the trust gap widened and the campaign for access and benefit sharing over genetic resources intensified (Parry 2004).

Some analysts even argue that the global seed banks were a convenient creation of colonial powers to ensure uninterrupted funnelling of genetic resources from colonial outposts, which could not be guaranteed in light of the transition to independence in those countries (Mgbeoji 2006).[2] Thus, the advancement of ag-biotech contributed to the coalescing of the debate over "biopiracy," IP, and access and benefit sharing.[3] One of the areas of this debate that has become increasingly urgent is plant genetic resources in global ex situ seed banks, which are reservoirs for some of the world's most valuable genetic resources. This debate hastened the signing of the ITPGRFA in 2001. One of the treaty's goals is to give effect to the idea of "farmers' rights." Those rights are emphasized in recognition of "the enormous contribution that the local and indigenous communities and farmers of all regions of the world ... make for the conservation and development of plant genetic resources, which constitute the basis of food and agriculture production throughout the world" (ITPGRFA, Article 9.1). The notion of farmers' rights underscores four important points: (1) the IP system does not reward the contributions of traditional farmers; (2) R&D innovation in ag-biotech and agricultural genomics does not occur in isolation from traditional knowledge (TK) and innovation in agricultural practices; (3) traditional farmers are critical agents and partners in the conservation of plant genetic resources for food and agriculture, which are fundamental to R&D in agricultural genomics; and (4) equitable access and benefit sharing is an imperative in the emerging agricultural innovation environment.

The Pre-ITPGRFA IP Landscape in Agriculture

With regard to the first of the four points underscored by the notion of farmers' rights, even though the CGIAR system (to which I shall return later) encourages the relaxation of IP rules over plant genetic resources held in trust, the prevailing global IP environment in the second half of the twentieth century favoured tightened IP protection. For instance, most technologically advanced European countries and others, including Japan, the United States, and Canada, were already members of an elite club of plant-breeding countries under the aegis of the International Union for the Protection of New Varieties of Plants (UPOV). UPOV developed a dedicated regime of *sui generis* IP rights known as plant breeders' rights (PBRs) that, despite the cosmetic accommodation of "farmers' privilege," under some circumstances has the effect of preventing farmers from saving or re-planting seeds of protected plant varieties (Dutfield 2008). Even though most UPOV member countries had domestic equivalents of PBRs, UPOV filled the gaps in international and domestic patent laws that did not directly sanction patents for plant breeding or other forms of agricultural innovation, or on life forms. UPOV was designed to promote the role of plant breeders in agricultural innovation at the expense of farmers, and it created a wedge between the two groups (Borowiak 2004; Dutfield 2008). Such a schism between the two was traditionally nonexistent in many ILCs in the Global South. UPOV has continued to keep pace with advancements in the science of plant breeding, to the point that, ironically, its rules and standards of protection now mimic the patent standard. Perhaps more important, as a system it emboldened a proprietary-centred regime, albeit with restricted sharing (i.e., for breeders only) in agricultural innovation.

Meanwhile, the historical reluctance to extend patent protection to life forms in general, and agricultural innovation in particular, was dissolving. This was a result of the World Trade Organization's *Agreement on Trade-Related Aspects of Intellectual Property Rights* (TRIPS Agreement). The United States and its fellow industrialized countries in Europe, Japan, Canada, and Australia, which form the bulk of the UPOV membership, were the principal champions of the TRIPS Agreement. This agreement extends patent protection to inventions, products, and processes with limited exceptions, without discrimination, to all fields of technology. Article 27 (the so-called biotechnology clause), specifically provides for patent protection for "micro-organisms, and essential biological processes for the production of plants or animals." Perhaps more important for this discussion, it requires either patent protection or what it calls "effective *sui generis*" protection for plant varieties, or a combination of both (Article 27[3][b]) to be instituted by member countries. Article 27 represents a broad extension of the patent regime into agricultural innovation. In addition to patents, it also envisages alternative or *sui generis* forms of IP protection, which include, but are not necessarily limited to, the UPOV plant breeders' rights system. This radicalization of the IP landscape in agricultural innovation has implications not only for the conduct of R&D in agricultural genomics but also for fair and equitable access to the benefits of such R&D.

Membership in UPOV is voluntary.[4] Consequently, before the TRIPS Agreement, countries were under no obligation, save pursuant to bilateral or related agreements, to protect plant breeders' rights or to overtly promote the dichotomization of breeders and farmers. However, membership in the WTO automatically commits a party to all WTO constitutive agreements, including the TRIPS Agreement, requiring members to protect new plant varieties in domestic patent laws, through an alternative *sui generis* option, or both. What is more, under the WTO process, failure of a member state to comply with the TRIPS Agreement's minimum standards of IP protection could trigger serious trade sanctions. Both the UPOV Convention and the TRIPS Agreement designate the standard of IP protection in agricultural innovation. Developed countries sponsored both regimes (Sell 2003; Aoki and Luvai 2007) to optimize their techno-scientific advantage and their head start in agricultural genomics, plant breeding, and ag-biotech in general. It is in this context of pre-existing national commitments to tightened IP protection for agricultural innovation that I now evaluate the implementation experience of the ITPGRFA.

The ITPGRFA in Cognate Regimes

The ITPGRFA is not an isolated or exclusive agricultural treaty. Rather, it exists at the intersections of biodiversity conservation, access and benefit sharing with regard to plant genetic resources (Chiarolla 2008a), and protection of TK, specifically farmers' rights. Farmers' rights represented a counterbalance to the role of IP in agricultural innovation, especially with regard to IP's indifference to the mainly informal contributions of traditional farming communities. On biodiversity and access and benefit sharing, the ITPGRFA shares identical objectives with the *Convention on Biological Diversity* (CBD). Technically, however, the ITPGRFA is concerned with plant genetic resources relevant to food and agriculture under the more focused FAO framework, whereas the CBD is concerned with a broad range of genetic resources, including animals and plants for nonfood and non-agricultural purposes. Specifically, in Article 1.1, the treaty states its objectives as "the conservation and sustainable use of plant genetic resources for food and agriculture and the fair and equitable sharing of the benefits arising out of their use, *in harmony with the Convention on Biological Diversity*, for sustainable agriculture and food security" (emphasis added). Article 1.2 further provides: "These objectives will be attained by closely linking this Treaty to the Food and Agriculture Organization of the United Nations and to the Convention on Biological Diversity."

By virtue of this unmistakable link between the CBD and the ITPGRFA, both instruments are crucial to the sustainable management of plant genetic resources for food and agriculture, and indeed all genetic resources. Consequently, familiarity with their objectives and operational mechanisms is vital for researchers and diverse stakeholders in agricultural genomics. These two instruments are premised on the understanding that plant genetic resources for food and agriculture are the core raw materials for innovations in agricultural genomics and other forms of ag-biotech, and for global environmental sustainability more generally. Both also recognize the role of ILCs and their knowledge systems in the conservation of plant genetic resources. In addition, they are committed to the entrenchment of a regime of equitable access to such resources (by agricultural R&D interests) and benefit sharing of resulting innovations (by broader segments of society, especially members of ILCs, whose stewardship of these resources through traditional knowledge and other traditional practices are globally recognized). In an important way, both regimes underscore the pluralism of the epistemic order in agricultural innovation and various forms of bio-cultural innovation. As I have noted earlier, a strong and unbalanced IP regime constitutes a threat to the realization of socially beneficial outcomes from agricultural innovation as embedded in the vision of both the ITPGRFA and the CBD. In essence, the marriage of the ITPGRFA, the CBD, and the access and benefit-sharing process helps to amplify policy expectations for the conduct of public-oriented and social impact research in agricultural genomics.

Core Treaty Provisions

Notably, there is no mention of farmers' rights in the objectives of the ITPGRFA outlined above. However, Article 9 of the treaty is the focus of its implementation strategy on farmers' rights, linking the concept of such rights to all of the treaty's objectives. As already noted, farmers' rights aim at recognizing the role of farmers, in ILCs and elsewhere, in the stewardship of plant genetic resources for food and agriculture, which are at the core of the global food supply. Under the treaty, securing farmers' rights involves the following: (1) protecting TK relevant to plant genetic resources for food and agriculture and the right to equitable participation of farmers in the sharing of benefits from the utilization of such resources (Articles 9.2[a] and 9.2[b]); (2) ensuring participation of farmers in decisions that affect them (Article 9.2[c]); and (3) supporting pre-existing rights that farmers have to "save, use, exchange and sell farm-saved seed/propagating material" (Article 9.3). It is important to indicate here that "utilization of plant genetic resources for food and agriculture" is a broad term that encapsulates most R&D in agricultural genomics and broader ag-biotech research, including socially beneficial applications or outcomes.

As part of the treaty's implementation framework, Article 10 creates a multilateral system of access and benefit sharing to regulate dealings in sixty-four plant genetic resources listed in Annex 1 (including subsequent accessions), some of which are under the control of the CGIAR International Agricultural Research Centres' ex situ global seed banks. The treaty also encourages all stakeholders – including national seed banks, individuals, ILCs, and corporations in possession of the plant genetic resources listed in Annex 1 – to incorporate such resources into the multilateral system. Between 2004 and 2009, a total of 1,170 such accessions, including over 1.3 million samples, have been made available to the multilateral system (Bhatti 2009). In essence, the system is a global mechanism through which the ITPGRFA and cognate regimes aspire to ensure that benefits of agricultural R&D and innovations in agricultural genomics and ag-biotech at large are managed under an access and benefit-sharing framework for the advancement of farmers' rights, sustainable agricultural R&D, and conservation of plant genetic resources for food and agriculture. In their practical detail, the implementation of these objectives has the potential to facilitate socially beneficial outcomes of R&D in agricultural genomics and other forms of ag-biotech.

The treaty makes detailed provisions for equitable sharing of benefits arising from the utilization of plant genetic resources for food and agriculture in the multilateral system. They include facilitated exchange of information, access to and general transfer of technologies, and capacity building relevant to the use of improved varieties and genetic materials from the multilateral system, in accordance with the priorities of the FAO's Global Plan of Action on plant genetic resources for food and agriculture.[5] Under Article 17, the treaty establishes a global information system for the exchange of information on plant genetic resources for food and agriculture among contracting parties. A constant theme in the treaty is the facilitated transfer of technologies relevant to both plant genetic resources for food, agriculture, and conservation, including those protected by IP, for the benefits of farmers in developing countries, least-developed countries, and countries with economies in transition. Yet, illustrative of its characteristic textual conflicts and ambiguities surrounding IP, the treaty drops its generally moderate tone and insists that access and transfer of technologies under its framework "shall be provided on terms which recognize and are consistent with the *adequate and effective* protection of intellectual property rights" (Article 12.2[b][iii]; emphasis added). The treaty also encourages creative strategies to cultivate private and public partnerships in developing countries and other forms of collaboration in agricultural R&D for plant genetic resources for food and agriculture under the multilateral system, including flexible forms of benefit sharing.

In an unequivocal manner, Article 13.3 states (emphasis added) that

> benefits arising from the use of plant genetic resources for food and agriculture that are shared under the Multilateral System should flow *primarily, directly and indirectly, to farmers in all countries, especially in developing countries, and countries with economies in transition, who conserve and sustainably utilize plant genetic resources for food and agriculture.*

Six years into the treaty's implementation experience, the focus on the operationalization of farmers' rights in developing countries is evident. As noted in Article 18, the treaty's financial provisions, the impact of the treaty on the targeted beneficiaries will depend mostly on the extent to which developed countries make contributions to the Benefit Sharing Fund (BSF). The fund, which was established in 2008 by the treaty's Governing Body as part of its funding strategy, is a milestone in the implementation of the treaty.

Adopted in 2004 by the Governing Body pursuant to Article 18, the funding strategy for the implementation of the treaty is premised on principles geared toward promoting transparency, effectiveness, and efficiency for the generation and disbursement or availability of funds, and the overall management of financial resources under the auspices of the Governing Body. Within the purview of the funding strategy are innovative designs for creative resource mobilization to support the treaty objectives. This also includes a focus on how best to ensure that relevant resources outside the control of the Governing Body further these objectives. The nucleus of the funding strategy is the BSF, the collective account for funding agricultural R&D for plant genetic resources for food and agriculture pursuant to the multilateral system administered under the treaty framework.

Financial contributions to the BSF are both mandatory and voluntary. Under the mandatory stream, pursuant to Article 13.2, recipients who prefer to commercialize exclusively – by IP or by other means – products of agricultural R&D derived from genetic material accessed from the multilateral system (instead of making them freely available) are required to pay "an equitable share of the benefits arising from the commercialization of that product." Such payment is to be made into a trust account, which the Governing Body has since established (in the form of the BSF) in accordance with Article 19.3(f) of the treaty. It is also instructive that the mandatory contribution requirement is a component of the Standard Material Transfer Agreement (SMTA), which articulates the fine details and terms for exchange of genetic resources in the multilateral system as prescribed under Article 12.4. The SMTA mechanism is an idea that developed under the CBD framework and was adapted by the Governing Body into the treaty. Under the treaty, it provides the contractual framework for ensuring that dealings in plant genetic resources for food and agriculture from the multilateral system by actors, including donors and recipients, comply with access and benefit sharing and other social interest objectives of the ITPGRFA (Chiarolla 2008b, providing operational details of the SMTA under the ITPGRFA; De Jonge 2011, providing critical perspectives on the amorphous concept of equitable access and benefit sharing in international treaties).

With regard to voluntary contributions, the treaty encourages financial donations to the BSF from a variety of sources, especially parties in developed countries, corporations or private sector actors, international or national organizations, individuals, and so on. An earlier attempt to create a similar optional funding arrangement under the IUPGRFA system failed because it was shunned by developed countries. As noted previously, the ITPGRFA recognizes that its success depends significantly on the extent to which wealthy member states are willing and able to support the funding regime. So far, only a few such countries, including Norway, Italy, Switzerland, Spain, Canada, Australia, Ireland, and Indonesia, have provided voluntary financial support to the BSF. The Governing Body is currently at a fairly early stage of exploring the potential of a sustainable and robust BSF capable of advancing treaty objectives. The Governing Body has continued to collaborate with more experienced organizations, such as the Global Crop Diversity Trust, and relevant international and intergovernmental partners in order to shore up competence in resource mobilization (Report of the Third Session of the Governing Body of the ITPGRFA 2009, paras. 31–32).

In addition to the above-listed parties to the treaty, the FAO is another source of funding support for the ITPGRFA. For the five-year period from 2009 to 2014, the Governing Body projected that US$116 million will accrue to the BSF (Bhatti 2009).

Project Funding under the ITPGRFA

So far, the Governing Body has developed a robust and transparent process for evaluating and awarding research project funds under the BSF. Through the Bureau (the treaty's administrative office), it has successfully awarded research funds for three project cycles in several countries, all of which are developing countries, least-developed countries, and countries with economies in transition.[6] Some of these projects target multiple countries as potential beneficiaries outside the principal or primary submission countries.[7] The projects covered under the two pioneer funding cycles were selected as priority projects by the Governing Body on the basis of the FAO's Global Plan of Action. Priorities include agricultural R&D relevant to sustainable conservation; capacity building for local farmers and other actors with regard to agricultural genomics skills, including crop adaptation for climate change; and on-farm conservation strategies. The latter two priorities were the specific focus of the 2011 awards. The range of local crop varieties involved and the diversity of institutional hosts of funded research projects under the BSF reflect a focused determination to promote capacity building in agricultural R&D, access and benefit sharing, and other socially beneficial outcomes of agricultural genomics and innovation for populations most in need.[8]

The ITPGRFA and CGIAR Systems

In addition to the projects being funded under the BSF, an important aspect of the implementation experience of the ITPGRFA is the assertion of its jurisdiction over ex situ collections of plant genetic resources for food and agriculture held by the CGIAR's federating International Agricultural Research Centres (IARCs). The CGIAR is an institutional network of private and public sector interests in the pursuit of agricultural R&D. Until not long ago, the CGIAR held the largest single collection of plant genetic materials, comprising 650,000 accessions (about 10 percent of the world's collections) (Shaw 2009, 161; Oguamanam 2011, 289), mainly from developing countries, least-developed countries, and countries with economies in transition. The plant genetic resources for food and agriculture under the CGIAR's auspices are held in fifteen IARCs spread across different regions of the world. Over the years, these IARCs have been sites of cutting-edge collaborative research in agricultural genomics and various forms of ag-biotech. As mentioned earlier, the active role of private sector partners in the activities of the CGIAR, and the opaque manner in which IP issues were dealt with in the CGIAR system, created a crisis of confidence between developing countries and their developed counterparts. This state of affairs highlighted the urgent need for transparency, accountability, and access and benefit sharing over the ex situ plant genetic resources in the CGIAR system. In principle, that system encourages relaxation of IP rights with regard to the utilization of benefits of plant genetic resources held in trust. However, the fifteen CGIAR IARCs pursue different interests and operate independently and without a legal charter (Marden and Godfrey 2012). Save for a 1994 agreement between the FAO and the IARCs governing dealings with in-trust plant genetic resources (Chiarolla 2008b), there was no clear juridical or enforcement framework in international law for securing strict compliance or for holding the CGIAR accountable to the trust in which it held the plant genetic resources in the ex situ seed banks.[9] This is a gap that the ITPGRFA seeks to mitigate.

In Article 15, the ITPGRFA asserts jurisdiction over all the plant genetic resources listed in Annex 1, which are held in the CGIAR's ex situ seed banks, and subjects them to the multilateral system established under Part IV of the treaty. Plant genetic resources that are part of Annex 1 and were collected before the treaty came into force, are dealt with in accordance with the Material Transfer Agreements already in operation between the CGIAR's IARCs and the FAO. As part of its implementation experience, the Governing Body abandoned the idea of using two Material Transfer Agreements for annexed and non-annexed plant genetic resources. Rather, it preferred to use the SMTA developed for Annex 1 plant genetic resources, which are integral to the multilateral system, and endorsed the addition of interpretative footnotes to relevant portions of the SMTA, indicating that the latter were applicable to non-annexed plant genetic resources collected before the treaty came into force (Report of the Second Session of the Governing Body of the ITPGRFA 2007, para. 68).[10]

The gist of Article 15 of the ITPGRFA is the vesting of authority in the treaty's Governing Body over dealings in plant genetic resources held by the IARCs and subject to the provisions of the treaty. Also under Article 15, IARCs are required to sign agreements with the Governing Body, committing them to comply with the SMTA and other terms for dealings with plant genetic resources for food and agriculture acquired from contracting parties. In all, the treaty is designed to ensure that the IARCs' dealings with plant genetic resources for food and agriculture from treaty parties are conducted under standard norms of accountability, transparency, effectiveness, and efficiency through the SMTA and other access and benefit-sharing protocols as developed under the multilateral system and treaty implementation work programs. As a matter of priority, the first session of the Governing Body held in Madrid considered draft agreements (which have since been adopted) between the Governing Body and the IARCs and other international institutions in the spirit of Article 15 of the treaty (Draft Agreements between the Governing Body and the IARCs of the CGIAR 2006).[11] Finally, through its secretary, the treaty has an oversight role with regard to treaty-relevant activities of organizations dealing with plant genetic resources for food and agriculture, besides providing those organizations with a range of technical support measures.

Next to the CBD, the CGIAR is perhaps the most important ITPGRFA partner for the realization of the treaty's objectives, especially in the non-monetary aspects of the benefit-sharing scheme. The entry into force of the ITPGRFA and its continuing implementation program have brought the two together in very meaningful ways, in terms of forging closer collaboration and sustaining ongoing partnerships for a more socially beneficial outlook with regard to equitable access to and benefit sharing in agricultural R&D among stakeholders. Before the coming into force of the ITPGRFA and its program implementation initiatives, the CGIAR process had a number of programs under its strategic research framework aimed at promoting capacity building and grassroots adaptation and development of local expertise and external innovations in agricultural genomics, devoid of IP constraints. One such example is the HarvestPlus (HP) program. The HP program involves the use of molecular biology and other agricultural genomic methods for bio-fortification – that is, to boost the nutritional or even medicinal value of foods for functional ends. The HP program is a practical reification and localization of R&D innovation in agricultural genomics. It targets key crops (including beans, cassava, sweet potato, rice, maize, and wheat) selected not on the basis of their commercial profile but because of their endemic value and association with the food culture of local peoples (Oguamanam 2011, 288–92).

So far, the CGIAR continues to collaborate with the ITPGRFA Governing Body through the development of treaty standard-setting norms and exploration of strategic pathways for impact and policy translation. A CGIAR representative serves on treaty advisory technical committees, notably, those on the SMTA and on the multilateral system. Furthermore, the IARCs have remained engaged with the treaty through the development and implementation of the latter's new information technology tool, Easy-SMTA, which is now an operative online mechanism servicing the multilateral system in terms of access and benefit sharing.[12] Easy-SMTA facilitates the development and compilation of SMTAs in the treaty's official languages and the reporting of SMTAs concluded under the treaty process, which is now an obligatory requirement for the IARCs and, arguably, for dealers in plant genetic resources for food and agriculture in treaty member states.

In 2006, the FAO signed (on behalf of the Governing Body) agreements with eleven IARCs in possession of in-trust plant genetic resources for food and agriculture pursuant to Article 15 of the treaty. In addition to that agreement, the setting up of the Benefit Sharing Fund, the implementation of harmonized SMTAs, and indeed the establishment of the multilateral system have collectively coalesced in the assumption or transfer of control and governance of global plant genetic resources for food and agriculture to the ITPGRFA regime. Under this new orientation, the issue of intellectual property in the CGIAR system remains obscure and in need of urgent elucidation, if not review.

CGIAR Reform and Principles for the Management of Intellectual Assets

The CGIAR embarked on a reform process in 2008. The reform resulted in a change of name to the "CGIAR Consortium" in April 2010, reflecting a streamlining of the IARCs and diverse stakeholders, including funders and researchers, for efficient and effective results. Reflecting the historical possessive interests of developed countries in the CGIAR, the consortium is hosted in Montpellier, France, while Washington, DC, serves as the host of the CGIAR Consortium Fund. The changes reflect a new commitment to harmonization of donor contact and centralized access to the CGIAR's International Agricultural Research Centres (IARCs). In addition, the changes represent a bid to streamline the consortium's strategy and research framework, harmonize the IARCs' research programs, and ensure operational transparency and delivery of measurable impacts.

Related to funding priorities under the Benefit Sharing Fund, the CGIAR reforms aim to strengthen individual IARCs in areas of factor endowment on identified global or regional issues, under the Challenge Programs. These programs are framework models for collaborative research that are aimed at exploring and exploiting complementarities, synergies, and collaboration among CGIAR IARCs and partners in order to break down institutional and disciplinary barriers to knowledge dissemination for effective impact (CGIAR 2013). Other core areas of interest include innovations at the intersection of water and food, specifically, reducing the pressure on the global water supply arising from agricultural activities, increasing micronutrients in staple food crops endemic to the rural poor under the HP program, and developing a new generation of agricultural crops responsive to changing agronomic needs and consumer demands. The details of these initiatives are outside the scope of this chapter.

With regard to intellectual property, the CGIAR Consortium and the Consortium Fund Council approved the CGIAR Principles and Guidelines on the Management of Intellectual Assets to guide the handling of intellectual asset (IA) issues in CGIAR research work from all funding sources (CGIAR 2012a).[13] This two-year experimental initiative came into effect in March 2012 and is expected to generate evidence and learning to assist in the design of a more permanent regime. Essentially, the principles are designed to fill the gaps in the new regime for the management of plant genetic resources for food and agriculture under the ITPGRFA as they relate to plant genetic resources for food and agriculture under development in the IARCs. Specifically, Article 6 provides for flexibility in dealing with derivatives under development, including conferring discretion for imposing additional conditions for their transfer. Article 8 of the SMTA vests the FAO with third-party beneficial and representative interests on behalf of the multilateral system (Chiarolla 2008b) in determinations or negotiations relating to such derivatives. Hinting at the gaps in IP and the inchoate nature of the access and benefit-sharing framework of the treaty, Emily Marden and R. Nelson Godfrey note that the extent to which the access and benefit-sharing programs of the treaty pursuant to the SMTAs "provide the necessary balance to proprietary rights in providing and preserving resources for innovation" (2012, 384) is not clear. In the words of the CGIAR Consortium (CGIAR 2012b, section 2):

> The Treaty, the SMTA and the Centres' agreements with the Governing Body of the Treaty leave considerable flexibility with respect to the conditions (additional to those in the SMTA) under which providers may transfer PGRFA [plant genetic resources for food and agriculture] under development. The CGIAR Principles guide the Centres on how they can exploit that flexibility. In this regard, the CGIAR IA [intellectual assets] Principles

are structured so as to be entirely in line with, and complementary to, the Centres' 2006 agreements with the Governing Body.

The CGIAR IA principles are significant for several reasons. First, they underscore the continuing relevance of IP with regard to publicly funded agricultural R&D and utilizations of plant genetic resources for food and agriculture in the dedicated pursuit of socially beneficial outcomes, including equitable access and benefit sharing and various overlapping objectives of cognate regimes such as the ITPGRFA, the CBD, and other inter-regime platforms. Second, they also reflect the need for delicate negotiations and calibrations of diverse interests at the heart of complex partnerships with private, public, and other hard-to-classify-stakeholders involved in the operations of the CGIAR Consortium, and indeed in the new landscape of R&D in agricultural genomics and innovation. For instance, in addition to over sixty-four governments and NGOs, fifteen research centres, over two hundred miscellaneous partner organizations (including national and regional agricultural research bodies), and high-profile private sectors and global charities, the CGIAR Consortium comprises four principal funders: the FAO, the International Fund for Agricultural Development, the United Nations Development Programme, and the World Bank. It operates in 150 locations and its R&D outcomes have impacts on two hundred countries globally. Third, in terms of timing, there is perhaps no more auspicious time during which to refocus on the issue of intellectual assets or intellectual property than when the treaty system is clearly in a learning curve and is therefore still malleable.

At the core of the CGIAR IA principles is the entitlement of the consortium and the centres to retain their pre-existing rights to intellectual assets (Principle 9.3). The IA principles inform the common operational framework of the consortium and comprise various agreements – including joint agreements, consortium performance agreements, program performance agreements, and incidental downstream agreements – concluded before or after the principles came into effect. They also apply to renewals and extensions of agreements, including those involving funding with third parties after the agreement came into effect. The IA principles are premised on the understanding that the CGIAR IARCs' agricultural R&D outputs have the status of international public goods, which have to be sustained by "innovatively designed, carefully managed and diligently monitored incentives" (Principle 2), under creative partnerships that support widespread diffusion of research results. This is done in order to achieve "the maximum possible access, scale and scope of impact and sharing of benefits to the advantage of the poor, especially farmers in developing countries" (Principle 1). Logically, the CGIAR IA principles also declare the consortium's support for farmers' rights pursuant to the ITPGRFA (Principle 3).

The CGIAR IA principles permit flexible uses of IP by the consortium and its third-party collaborators on a prudent, contingent, and strategic basis that accommodates, among other considerations, limited use or limited exclusivity in terms of scope of rights and geographical segmentation, as well as research and emergency exemptions. The application of these concessions is subject to a rigorous reporting and accounting geared toward transparency and scrutiny to determine whether such concessions are justified or necessary to enhance the CGIARs' intellectual assets, advance the scope of their impact on beneficiaries, and further the overall CGIAR vision. The CGIAR IA principles vest in the consortium and centres the discretion to develop intellectual assets with downstream partners and other stakeholders on terms consistent with the CGIAR vision. They also give the centres responsibility for the management of their intellectual assets.

The CGIAR IA principles recognize that since the consortium does not have a monopoly on global plant genetic resources for food and agriculture or agricultural R&D generally, there is a need to reach partners who operate outside the public good matrix. According to the consortium's general counsel, "reaching intended beneficiaries [with the products of CGIAR research] sometimes requires Centers to take up IP protection, grant exclusive licences and/or access third party IAs that [may] have downstream restriction" (Perset 2012). Under the complex framework in which the consortium is organized and operates, the IARCs are able, pursuant to the CGIAR IA principles, to grant exclusive licences toward commercialization of its intellectual assets in circumstances where exclusivity is imperative so long as it is limited "in duration, and/or field of use and if the IAs are available in all countries *for research* by public organizations ... and in the event of national or regional food security emergency" (Perset 2012; emphasis added). Overall, the CGIAR IA principles represent a delicate attempt at negotiating socially beneficial open and nonexclusive global access to the CGIAR Consortium's intellectual assets by the poor and those in need, in balance with the proprietary interests of private sector stakeholders. The CGIAR IA principles, however, do not sanction extension of IP protection to in-trust germplasm in the consortium's possession.

The CGIAR IA principles were formally introduced to and welcomed by the treaty in October 2012. At present, they are part of the ongoing collaborative efforts between the consortium and the treaty. The Treaty Secretariat has expressed a willingness to advise the consortium on the IA principles, if need be, and to continue to monitor the process of its implementation in liaison with the consortium office (CGIAR 2012b).

Ambiguities and Mixed Signals on Intellectual Property

Evidently, the implementation of the treaty poses complex regulatory and IP challenges. With regard to the latter, given the fact that both the treaty implementation and the recent reforms at the CGIAR are concurrently on a learning curve, the next two years – of test-running the IA principles and further deepening the global governance of plant genetic resources for food and agriculture under the treaty – will illuminate emergent IP challenges under the evolving framework between the two. So far, partly because of some gap-filling details in the SMTA, not much attention has been paid to the practical details and interpretive ramifications of the double-speak on IP that characterizes the treaty text. For instance, as I indicated above, despite the overall orientation of the treaty toward fair and equitable access to plant genetic resources for food and agriculture and the benefits of the agricultural R&D innovation, it insists on *"adequate and effective* protection of intellectual property rights" (Article 12.2[b][iii]; emphasis added). The extent to which this language reflects balance in the applications of IP under the treaty framework is debatable.

Perhaps most symbolically, Article 12.3(d) of the treaty bars IP rights claims over plant genetic resources for food and agriculture, including their genetic parts, but *only* in the form received from the multilateral system. In many quarters, the provision amounts to a bounced cheque with regard to the overall expectations of the treaty held by developing countries. A self-serving interpretation of this provision from an anti–farmers' rights perspective favours an understanding that IP rights can continue to be claimed by users of plant genetic resources from the multilateral system so long as they are able to merely manipulate or transform the germplasm from "the form received." The drafting history of that article and the resulting interpretive ambiguity reflects the tension between developed and developing countries in the treaty negotiations (Helfer 2002; Oguamanam 2006) as well as a delicate compromise between them. That article is the root of the controversy over what constitutes "derivatives" of plant genetic resources in the multilateral system, and whether such derivatives are subjects of the access and benefit-sharing requirement. In a way, that controversy would appear to have been somewhat resolved under Article 6 of the SMTA.[14]

Finally, under the treaty, the burden of implementing farmers' rights largely rests with national governments. As this chapter has demonstrated, the pre-existing domestic and international (e.g., UPOV, TRIPS, etc.) IP obligations of developed country parties to the treaty are at cross-purposes with the objectives of the treaty, especially in the areas of relaxation of IP protection in agricultural innovation and in promotion of farmers' rights. Despite the centrality of farmers' rights, the text of the treaty on that concept is indeterminate and bereft of tangible elements incidental to rights jurisprudence in law (Oguamanam 2006; Borowiak 2004). Instructively, while the treaty uses the language of rights in relation to farmers, albeit with equivocal jurisprudential ramification, the UPOV Convention adopts the language of privilege. As a consequence, there is significant room for confusion and ambiguity that makes it quite challenging for states to formulate domestic legislation that is sensitive to their obligations under the UPOV Convention and the ITPGRFA. Canada has recently processed an implementing legislation to the 1991 revision of the UPOV Convention through Bill C-18 (*Agriculture Growth Act*). Some stakeholders, especially segments of farmer interest groups, opposed the bill's use of "privilege" instead of "right" in relation to its accommodation of farmers' practice of seed saving.[15] All these demonstrate that the issue of farmers' rights is the flashpoint of IP and other nuanced proprietary regime complexes over plant genetic resources for food and agriculture.

Regulatory Complexity and Collaboration Imperatives

This chapter has illustrated the extent of overlap between the ITPGRFA objectives and the CBD as well as the CGIAR Consortium. An in-depth appraisal of the various jurisdictional overlaps over the subject matters of the treaty unveils a regulatory complex of unimaginable scope. The shared objectives of the CBD and the treaty engage biodiversity and conservation imperatives, since plant genetic resources for food and agriculture are crucial aspects of biodiversity. Both regimes also engage imperatives for the protection of traditional knowledge, which is at the heart of the CBD's implementation strategy. The treaty conceives of farmers' rights as an integral part of TK. Consequently, the treaty has a stake in terms of coordination and collaboration in the very robust and complex sites for negotiating and promoting biodiversity conservation and the TK and farmers' rights relevant thereto. These include various work programs and protocols of the CBD, notably the *Nagoya Protocol* on access and benefit sharing, as well as on-going negotiations at the World Intellectual Property Organization (WIPO) Intergovernmental Committee and the burgeoning international and domestic legal framework on access and benefit sharing.

The ITPGRFA Governing Body has developed a broad network of institutional collaborations through which it engages various global actors in food governance issues, including collaborative capacity building, financial support, information exchange, norm development, and management of jurisdictional overlap (Resolution 8/2011 of the Fourth Session of the Governing Body of the ITPGRFA). Its engagement with the Commission on Plant Genetic Resources for Food and Agriculture (CPGRFA) assists in creating internal awareness of the state of the world's plant genetic resources for food and agriculture in the treaty bureaucracy. Such commitment gives the treaty a foothold in the future review of the Global Plan of Action for the Conservation and Sustainable Utilization of Plant Genetic Resources for Food and Agriculture being undertaken by the CPGRFA. The treaty's robust relationship with the CBD cuts across various CBD work programs at various levels, including agricultural biodiversity, sustainable use of biodiversity, and biodiversity and climate change. Recently, its collaborative work with the CBD Secretariat on access and benefit sharing was repositioned around the emerging processes of the *Nagoya Protocol*, which came into effect in October 2014. On a broader scale, the Treaty Secretariat participates in the Liaison Group of Biodiversity-Related Conventions.[16] As well, the treaty encourages cooperation with UPOV, the World Health Organization, WIPO, and other relevant organizations. In essence, the treaty's proactive collaboration underscores the pre-existing regime complexity around its subject matter and the determination of the Governing Body to create awareness around the treaty even as these regimes continue to proliferate.

Today's global profile of hunger shows that 842 million people are chronically hungry. Seventy-five percent of the world's hungry or poor live in rural areas of the developing world and are engaged in agricultural production.[17] Agricultural R&D in genomics and broader biotechnologies has continued to increase since the discovery of recombinant DNA. Efficient management of global genetic resources has become an imperative if the world is to tackle the interconnected burdens of hunger, poverty, and disease that are rampant among the world's most vulnerable. An important segment of these vulnerable groups comprises local farmers who have immemorial traditional patrimony and stewardship of much of the world's global plant genetic resources for food and agriculture.

The ethos of equitable access and benefit sharing and other socially beneficial orientations with regard to agricultural innovation is truly something that is characterized by excessively complex regulatory and IP intervention. This chapter has mapped out the scope of these regulatory and IP dynamics with a focus on the implementation of one of the most far-reaching such initiatives so far, the ITPGRFA, in the context of the pre-existing framework under the new CGIAR Consortium. Aside from the identified regulatory complexity, which is an inherent aspect of plant genetic resources for food and agriculture as a subject matter at the intersection of multiple regimes and multiple stakeholders, this chapter also identified juridical weaknesss in the way IP is nested across these regimes. Perhaps more important, it is clear from the analysis that even though IP may act as an impediment for the optimal uptake of the benefits of agricultural R&D, IP rights can serve important and desirable goals even proactively in the context of public interest orientations. Essentially, this is so because of the melange of stakeholders and the complex layers of collaboration needed to secure optimally beneficial results.

The Governing Body's implementation of the treaty has transformed it into the only fully operational global instrument on access and benefit sharing with regard to plant genetic resources for food and agriculture (Bhatti 2009; Chiarolla 2008b). In so doing, the Governing Body has established an elaborate, transparent process in the way it deals with plant genetic resources for food and agriculture in both the annexed and non-annexed categories. So far, the impact of the ITPGRFA initiatives has coincided with important renewals at the CGIAR on several fronts – notably, for our purposes, with regard to the CGIAR IA principles. The extent to which the CGIAR IA principles can be operationalized in harmony with the treaty regime will be fundamental to new thinking around the role of IP as a potential facilitator of socially beneficial outcomes of agricultural R&D, in the spirit of the objectives of both the treaty and the CGIAR. Making sixty-four of the world's most important crops – which collectively represent 80 percent of all human food consumption – available for research on a streamlined and cost-efficient basis under the multilateral system is an incredible way of bringing together plural epistemic interactions in agricultural innovation. This is far more desirable than the winner-take-all approach that an unbalanced IP system could accomplish.

There are high expectations for the ITPGRFA thanks to what has already been achieved in the short time since its implementation. The Governing Body's ability to attract funds into its Benefit Sharing Fund is very important. To date, only a few developed countries have identified with the fund, even on a miserly scale that is hardly satisfactory. Given the global scope of the treaty and the mandate of the Governing Body, it is quite clear that funding is important to its success. The treaty budget does not even match a small fraction of departmental expenditures at the world's leading ag-biotech firms. There is a need to boost support for the treaty among all stakeholders. Aside from expanding funding for strategic priority R&D projects, there is also a need to build unique expertise and the logistical where-withal to equip the Treaty Secretariat for optimum delivery of the various technical support services it is required to provide.

As a reflection of both momentum in the ambitious implementation of the treaty and the enthusiasm of its supporters, a second High Level Round Table on the Treaty (HLRT) was held at the United Nations Conference on Sustainable Development in Rio de Janeiro on June 21, 2012 (Rio+20). It identified six priorities, or action plans, to be pursued by the treaty in collaboration with relevant stakeholders (Rio Six-Point Action Plan for the ITPGRFA 2012). Notably, they include development of a platform for technology transfer under a nonmonetary benefit-sharing framework; promotion of public/private partnerships for pre-breeding; and a centralized, broad negotiation to incorporate all plant genetic resources for food and agriculture into the treaty framework. Other priorities include the promotion of information on the values of underutilized species of national and regional significance to food security and sustainable development, and the generation of interest among stakeholders concerning the full implementation of the treaty and its ramifications with regard to such considerations as "food security, nutrition and resilience of agriculture systems, particularly in the context of climate change" (Rio Six-Point Action Plan for the ITPGRFA 2012, Action Point 5). Finally, the Governing Body is asked to consider the inclusion of additional crops in treaty Annex 1.

Issues around IP will remain front and centre in these ambitious future treaty projects. There are many more issues in need of clarity. A snapshot of such issues (in no particular order) includes diverse third-party interests at the intersection of public and private agricultural R&D, and the lingering issue of "derivatives" and associated interpretive ambiguities (Fowler et al. 2004). In addition, it is too early to determine the extent of the practical impact and uptake of research programs funded under the BSF by the ultimate target beneficiaries. It is yet unclear and requires dedicated exploration of the extent to which independent private (third) parties could advance nonmonetary benefit sharing under the treaty regime. Also, potential expansion of Annex 1 crops and forage plants would require broad negotiation and consultation with stakeholders, especially members of ILCs. It would be presumptuous to expect that these communities would patronize the multilateral system when its impact has yet to be felt.

Further, the *Nagoya Protocol* is a major and complementary access and benefit-sharing regime, and it is important to explore how that regime could exist side by side with the treaty and what lessons each may learn from the other (Chiarolla 2008b). Next to that, as more sophisticated advances continue in agricultural genomics and ag-biotech in general, the cumulative criteria for defining plant genetic resources for food and agriculture may be scientifically unsustainable, even to the point of blurring the jurisdictional demarcation between the treaty and the CBD. This demarcation is, for the most part, alien to ILC custodians of global plant genetic resources. And perhaps most important, the piecemeal, complex, and overlapping jurisdictional outlook of plant genetic resources for food and agriculture is such that staking uncritical faith on a single, seemingly overarching regime too early in time may be imprudent.

Overall, therefore, the means by which the treaty Governing Body implements the current experimental phase of the new CGIAR IA principles, as well as how it monitors transactions under the multilateral system as a whole, will set the tone with regard to IP in plant genetic resources for food and agriculture in all other systems, including the private sector. Without question, forging complex R&D partnerships in agricultural genomics at national and transnational levels requires actors, especially researchers, to be familiar with the fast-changing international governance landscape currently being radically revamped under the ITPGRFA and cognate regimes. At the centre of the emerging landscape are matters that transcend R&D in agricultural genomics. Most important, considerations pivotal to the conservation of germplasm and of sustainable agriculture – including fairness and equity over access and benefit sharing, intellectual property, traditional knowledge, imperatives for farmer-centred agriculture, and the scope of farmers' rights – have come to the fore. There is also a need for some reflection and foresight regarding potential future scenarios in the regulatory and IP outlook of the emerging landscape for the governance of plant genetic resources for food and agriculture.

Not all issues under the Rio+20 Six-Point Action Plan require immediate action. The impact of the treaty and its ramifications for public/private collaborators need to be appraised and fully understood by all stakeholders, not the least of which are agricultural R&D communities and members of ILCs. The HLRT process that generated the Six-Point Action Plan is gradually becoming a continual and important feature of the International Treaty on Plant Genetic Resources for Food and Agriculture experience. The third HLRT was held in Bandung, Indonesia, in July 2013 to shore up interest in the Benefit Sharing Fund among donors and top-level political influence wielders, and to appraise the policy contributions of the treaty toward "the global challenge of biodiversity, climate change and food security."[18] The HLRT will remain crucial in providing direction toward continual and critical evaluation of the treaty implementation, its consolidation in the short term, and its sustainability in the long term. It is necessary that the HLRT process be inclusive, responsive, and dynamic without compromising efficiency in the treaty implementation experience.

Acknowledgments
Thanks to Alessandra Toteda for exceptional research assistance, and to Elise Perset, CGIAR General Counsel, for our helpful conversations at The Hague in October 2012 and for follow-up correspondence during the course of preparing this chapter. I also thank Professor Jeremy De Beer, Emily Marden, and Nelson Godfrey for facilitating my attendance at the Conference on Changing Goals of Governance for Agricultural Genomics: An Examination of Current Regulatory and Intellectual Property Rights, in Vancouver, November 17–18, 2011. This chapter is based on a paper I delivered at that conference and I am grateful to all conference participants for their helpful feedback.

Notes
1 In this chapter, "agricultural genomics" is simply the application of genomics insights into the agricultural realm for agronomic innovation and the overall advancement of plant biology for crop improvement.
2 Although the CGIAR was established in 1971, the process of gaining strategic access to genetic resources from the Global South began well before its formal establishment, particularly through the activities of colonial powers and American interests.
3 The full extension of IP protection to life forms, including plant genetic resources for food and agriculture, pursuant to the *Agreement on Trade-Related Aspects of Intellectual Property* is generally perceived by analysts as a strategic empowerment of Western science and its industrial complex over dealings in plant genetic resources for food and agriculture at the centres of biodiversity. This resulted partly in rampant appropriation of pre-existing bio-cultural knowledge of indigenous and local communities concerning food, medicine, environmental management, etc.,

through the patent system (Mgbeoji 2006). This trend also led to a rise in the issuing of questionable "biopiracy" patents. "Biopiracy" is the term used to capture the underhanded and uncompensated appropriation of such bio-cultural knowledge of ILCs and their associated genetic resources by so-called second comers, mainly through the IP system (ibid.).

4 However, many developing countries are being recruited into UPOV – or at least into adopting the IP protection standards thereof – as part of TRIPS-plus components of bilateral free trade agreements with their developed-country counterparts.

5 Global Plans of Action are sets of strategic action plans for conservation and sustainable use of the world's genetic resources on a thematic basis in response to the FAO *Report on the State of the World's Plant Genetic Resources for Food and Agriculture* adopted by 150 countries in 1996 (http://www.fao.org/agriculture/crops/thematic -sitemap/theme/seeds-pgr/sow/en/) and its subsequent update in *The Second Report on the State of the World's Plant Genetic Resources for Food and Agriculture* in 2010 (http://www.fao.org/agriculture/seed/sow2/en/). So far, several such action plans have been adopted in a number of thematic areas, including plant genetic resources for food and agriculture (1996 and 2011 under the aegis of the Commission on Plant Genetic Resources for Food and Agriculture), on animal genetic resources (Global Plan of Action for Animal Genetic Resources 2007), and on forest genetic resources (2013).

6 The country beneficiaries under the first and second project cycles of the BSF include Cuba, Costa Rica, Egypt, India, Kenya, Morocco, Nicaragua, Guatemala, Peru, Senegal, Tanzania, and Uruguay. The others are Bhutan, Brazil, North Korea, Ethiopia, Guatemala, Indonesia, Jordan, Malawi, Nepal, Peru, Philippines, Sudan, Tunisia, and Zambia. For a list of funded projects and country beneficiaries under the third (2014) project cycle, see http://www.planttreaty.org/sites/default/files/ files/Third%20Call%20for%20Proposals-%20Projects%20approved%20for%20 funding-for%20web.pdf.

7 For example, the Nepalese project titled "Community Based Biodiversity Management for Climate Change Resilience" targets a total of eleven countries, most of them outside the Nepal region.

8 Including, e.g., maize/bean, potato germplasm, citrus varieties, finger millet, yam, nutmeg, pea, pepper, cassava, local durum, bread wheat, sorghum, etc.

9 Collections of plant germplasm within the IARCs under the auspices of the FAO were undertaken by virtue of an agreement between the IARCs and the FAO signed on October 26, 1994. During the transition to the ITPGRFA, those materials remained with the FAO, subject to the Material Transfer Agreement. However, that agreement was subsequently subsumed within the SMTA under the treaty, and subjected to the requirements of Article 15 of the treaty, including reporting requirements, inspections, and monitoring under the treaty system, exemption of in situ suppliers or origin countries from signing Material Transfer Agreements when they receive samples from the IARCs, etc.

10 See also para. 39 of the Report of the Third Session of the Governing Body of the ITPGRFA. The IARCs began using the SMTAs for non-annexed plant genetic resources on February 1, 2008.

11 Aside from IARCs, the Governing Body is required to establish treaty-compliant agreements with other relevant international institutions dealing with plant genetic resources for food and agriculture.

12 Easy-SMTA was launched by the Treaty Secretariat on February 17, 2012 (Secretariat of the International Treaty on Plant Genetic Resources for Food and Agriculture 2012).

13 In this context, "intellectual assets" includes and transcends IP rights and embraces both rights traditionally covered by IP and those arising from R&D activities in the IARCs.

14 Claudio Chiarolla (2008b, 14) argues that even though the SMTA does not define derivatives or what amounts to misappropriation of resources in the multilateral system, the "incorporation requirement both in the definition of 'Product' and in Article 6.7 of the SMTA" erases any doubt that any product from the multilateral system whose commercialization triggers benefit sharing is a derivative.

15 One of the most vocal opponents of Bill C-18 is Canada's National Farmers Union (NFU). Through its "save our seeds" campaigns, among other things, the union opposes the use of "privilege" to designate or accommodate farmers' traditional practice of seed saving and exchange under the bill. For details of the NFU's opposition to Bill C-18, see National Farmers Union, "Stop Bill C-18" at http://www.nfu.ca/issue/stop-bill-c-18.

16 Formed by decision VII/27 of the CBD Conference of the Parties, this group comprises the heads of six secretariats of biodiversity-related conventions: *UN Framework Convention on Climate Change* (UNFCCC); *UN Convention to Combat Desertification* (UNCCD); *Convention on Biological Diversity* (CBD); *Convention on International Trade in Endangered Species of Wild Fauna and Flora* (CITES); *Convention on Wetlands* (Ramsar Convention); *Convention on Migratory Species* (CMS); and the *World Heritage Convention*. The group is committed to ensuring synergy in the conduct of their respective mandates. For an overview of the group, see Biodiversity Liaison Group 2013.

17 See Food and Agriculture Organization 2013. Of the 842 million hungry people, 15.7 million live in the developed world while the rest live in the developing regions.

18 See ITPGRFA, "3rd High Level Round Table on the ... (IT-PGRFA) 2 July 2013 – Bandung (Indonesia): Concept Note," at http://goo.gl/f1MK1r.

References

Legislation, Treaties, and Agreements

Agreement on Trade-Related Aspects of Intellectual Property Rights, April 15, 1994, *Marrakesh Agreement Establishing the World Trade Organization*, Annex 1C, Legal Texts: The Results of the Uruguay Round of Multilateral Trade Negotiations 320 (1999), 1869 U.N.T.S. 299, 33 I.L.M. 1197 [TRIPS Agreement].

Bill C-18 (*Agriculture Growth Act*), October 16, 2013 (Canada). http://www.parl.gc.ca/LegisInfo/BillDetails.aspx?Language=E&Mode=1&billId=6373658.

Convention on Biological Diversity, June 5, 1992, 1760 U.N.T.S. 79, U.N. Doc. ST/DPI/1307, 31 I.L.M. 818 (entered into force December 29, 1993) [CBD].

Draft Agreements between the Governing Body and the IARCs of the CGIAR, and other relevant international institutions, June 12–16, 2006, IT/GB-1/06/9. http://goo.gl/8v3Df.

Global Plan of Action for Animal Genetic Resources, adopted by the International Technical Conference on Animal Genetic Resources for Food and Agriculture on September 7, 2007, ITC-AnGR/07/REP. ftp://ftp.fao.org/docrep/fao/010/a1404e/a1404e00.pdf.

International Convention for the Protection of New Varieties of Plants, December 2, 1961, 815 U.N.T.S. 89 (as revised at Geneva on November 10, 1972, on October 23, 1978, and on March 19, 1991) [UPOV Convention].

International Treaty on Plant Genetic Resources for Food and Agriculture, adopted November 3, 2001. S. Treaty Doc. No. 110-19 [ITPGRFA]. ftp://ftp.fao.org/docrep/fao/007/ae040e/ae040e00.pdf.

International Undertaking on Plant Genetic Resources for Food and Agriculture, adopted November 23, 1983, FAO Res. 8/83 [IUPGRFA]. http://apps3.fao.org/wiews/docs/Resolution_8_83.pdf.

Nagoya Protocol on Access to Genetic Resources and the Fair and Equitable Sharing of Benefits Arising from Their Utilization to the Convention on Biological Diversity, Conference of the Parties to the Convention on Biological Diversity, UNEP/CBD/COP/DEC/X/1 of October 29, 2010. [Nagoya Protocol].

Report of the Second Session of the Governing Body of the ITPGRFA, October 29 – November 2, 2007, IT/GB-2/07/Report. http://www.planttreaty.org/sites/default/files/gb2repe.pdf.

Report of the Third Session of the Governing Body of the ITPGRFA, June 1–9, 2009, IT/GB-3/09/Report. http://www.planttreaty.org/sites/default/files/gb3repe.pdf.

Resolution 8/2011 of the Fourth Session of the Governing Body of the ITPGRFA: Cooperation with Other Bodies and International Organizations, Including with IARCs of the CGIAR and Other International Institutions That Signed Agreements under Article 15, March 14–18, 2011.

Rio Six-Point Action Plan for the ITPGRFA, adopted by consensus at the second High Level Round Table on the ITPGRFA on the occasion of the United Nations Conference on Sustainable Development, Rio de Janeiro, Brazil, June 21, 2012. http://goo.gl/lyqPp.

Second Global Plan of Action for Plant Genetic Resources for Food and Agriculture, adopted by the FAO Council, November 29, 2011, C.L. 143/17. http://www.fao.org/docrep/015/i2624e/i2624e00.htm.

Secondary Sources

Aoki, Keith, and Kennedy Luvai. 2007. "Reclaiming 'Common Heritage' Treatment in International Plant Genetic Resources Regime Complex." *Michigan State Law Review* 35: 35–70.

Bhatti, Shakeel. 2009. "Technology Transfer Aspects of the Multilateral System of the ITPGRFA." Presentation at Food and Agriculture Organization International Technical Conference on Agricultural Biotechnologies in Developing Countries, Guadalajara, Mexico, March 3, 2009. http://goo.gl/6CzuF.

Biodiversity Liaison Group. 2013. "Liaison Group of Biodiversity-related Conventions." Convention on Biological Diversity, https://www.cbd.int/blg/.

Borowiak, Craig. 2004. "Farmers' Rights: Intellectual Property Regimes and the Struggle over Seeds." *Politics and Society* 32 (4): 511–43. http://dx.doi.org/10.1177/0032329204269979.

CGIAR (Consultative Group on International Agricultural Research). 2012a. "CGIAR Principles on the Management of Intellectual Assets." CGIAR Fund, http://www.cgiarfund.org/sites/cgiarfund.org/files/Documents/PDF/fc7_cgiar_ia_principles_inclusion_COF_Feb16_2012.pdf.

–. 2012b. "Implications of the CGIAR Policy Documents on the Management of Intellectual Assets for the implementation of the Multilateral System by the CGIAR Centres." Annex to IT/AC-SMTA-MLS 4/12/3.

–. 2013. "CGIAR Challenge Programs." CGIAR, http://www.cgiar.org/our-research/challenge-programs/.

Chiarolla, Claudio. 2008a. "The Question of Minimum Standard of ABS under the CBD International Regime: Lessons from the International Treaty on Plant Genetic Resources for Food and Agriculture." UNU-IAS Working Paper No. 156. http://www.ias.unu.edu/resource_centre/156%20Claudio%20Chiarolla.pdf.

–. 2008b. "Plant Patenting, Benefit Sharing and the Law Applicable to the Food and Agricultural Organization Standard Material Transfer Agreement." *Journal of World Intellectual Property* 11 (1): 1–28. http://dx.doi.org/10.1111/j.1747-1796.2007.00332.x.

De Jonge, Bram. 2011. "What Is Fair and Equitable Benefit Sharing?" *Journal of Agricultural and Environmental Ethics* 24 (2): 127–46. http://dx.doi.org/10.1007/s10806-010-9249-3.

Dutfield, Graham. 2008. "Turning Plant Variety into Intellectual Property: The UPOV Convention." In *The Future Control of Food: A Guide to International Negotiations and Rules on Intellectual Property, Biodiversity and Food Security,* ed. Geoff Tansey and Tasmin Rajotte, 27–47. London: Earthscan.

Evenson, R.E., V. Santaniello, and D. Zilberman, eds. 2002. *Economic and Social Issues in Agricultural Biotechnology.* New York: CABI. http://dx.doi.org/10.1079/9780851996189.0000.

Food and Agriculture Organization. 2013. "The State of Food Insecurity in the World: The Multiple Dimensions of Food Security." FAO Corporate Document Repository, http://www.fao.org/docrep/018/i3434e/i3434e00.htm.

Fowler, Cary, Geoffrey Hawtin, Rodomiro Ortiz, et al. 2004. "The Question of Derivatives: Promoting Use and Ensuring Availability of Non-Proprietary Plant Genetic Resources." *Journal of World Intellectual Property* 7 (5): 641–63. http://dx.doi.org/10.1111/j.1747-1796.2004.tb00222.x.

Global Plan of Action. 2013. "A Portal for Plant Genetic Resources for Food and Agriculture." Global Plan of Action, http://www.globalplanofaction.org/.

Helfer, R. Laurence. 2002. "Intellectual Property Rights in Plant Varieties: An Overview with Options for National Governments." FAO Legal Paper Online No. 31. http://goo.gl/auVsZ.

Kloppenburg, R. Jack, Jr. 1988. *First the Seed: The Political Economy of Plant Biotechnology, 1492–2000.* New York: Cambridge University Press.

Marden, Emily, and R. Nelson Godfrey. 2012. "Intellectual Property and Sharing Regimes in Agricultural Genomics: Finding the Right Balance for Innovation." *Drake Journal of Agricultural Law* 17 (2): 369–94.

Mgbeoji, Ikechi. 2006. *Global Biopiracy: Patents, Plants and Indigenous Knowledge.* Vancouver: UBC Press.

Oguamanam, Chidi. 2006. "Intellectual Property in Plant Genetic Resources: Farmers' Rights and Food Security in Indigenous and Local Communities." *Drake Journal of Agricultural Law* 11 (3): 273–305.

–. 2007. "Tension on the Farm Fields: The Death of Traditional Agriculture?" *Bulletin of Science, Technology and Society* 27 (4): 260–73. http://dx.doi.org/10. 1177/0270467607300638.

–. 2011. "Toward a Constructive Engagement: Agricultural Biotechnology as a Public Health Incentive in Less-Developed Countries." *Journal of Food Law and Policy* 17 (2): 257–96.

Parry, Bronwyn. 2004. *Trading the Genome: Investigating the Commodification of Bioinformation.* New York: Columbia University Press.

Perset, Elise. 2012. "Socially Responsible Management of Intellectual Asset in the CGIAR." Presentation at Workshop on Socially Responsible Licensing at the Hague Institute for Global Justice, October 9, 2012.

Secretariat of the International Treaty on Plant Genetic Resources for Food and Agriculture. 2012. Notification: Launch of Easy-SMTA. PL 40/31 NCP GB5 MLS SMTA, February 17. http://goo.gl/h1ZTm.

Sell, Susan K. 2003. *Private Power, Public Law and the Globalization of Intellectual Property Law.* Cambridge: Cambridge University Press. http://dx.doi.org/10. 1017/CBO9780511491665.

Shaw, D. John. 2009. *Global Food and Agriculture Institutions.* New York: Routledge.

Shiva, Vandana. 2000. *Stolen Harvest: The Hijacking of Global Food Supply.* Cambridge, MA: South End Press.

6

Intellectual Property Management and Legitimization Processes in International and Controversial Environments

JEREMY HALL, STELVIA MATOS, AND VERNON BACHOR

For many businesses in nearly all industries, knowledge is considered a principal economic asset and has become incorporated into the fabric of corporate strategy (Hanel 2006). As a result, both private and public institutions have recognized intellectual property (IP) as a valuable asset that requires protection and management. However, most contemporary research on IP management is based on a relatively narrow range of industrial settings, such as pharmaceuticals, chemicals, and electronics, and located in the relatively modern economies of developed countries with strong institutional support. Consequently, little is known about how such IP management research applies to different industrial settings and more complex environments where institutions are weak and there is social controversy about the technology. Such oversight has important implications, especially in industries with widespread social impacts such as agriculture, and particularly for those in developing and emerging economies because many poor, marginalized farmers in such countries may be adversely affected by modern IP management approaches. A failure to consider the nuances of differing industry characteristics, stakeholder views, and international institutional structures may exacerbate controversy and thus hinder acceptance of technology.

This chapter focuses on managing IP in the agricultural sector in challenging international settings. The case of Monsanto and the controversy over transgenic technology is explored in the context of the expansion of the agricultural biotechnology industry into the emerging economies of Brazil and India. We begin by describing the purpose and limitations of IP protection. We then discuss how IP management approaches need to align with the institutional context, followed by a discussion on how IP management may affect legitimization processes, which are dependent on the uncertainties under which the technology is being developed. We then present our methodology and case, which is analyzed considering the challenges of IP management for agricultural biotechnology in different institutional settings. We conclude by discussing the implications of our work for IP management, and by discussing the relevance of our work to this volume and the concept of the Intellectual Property–Regulatory Complex explored here.

The Logic and Limitations of Intellectual Property Protection

A primary outcome of innovation is intellectual property, where every new product or service has a new idea or approach, new ways of solving problems, applying a theory, or thinking about the world. IP needs to be protected in order for inventors to have an incentive to recover any associated costs of creation (such as research and development expenditures) and to earn a profit, for example, by gaining a period of monopoly in which their idea cannot be copied without compensation (such as through licensing). This recognition of an individual's right to benefit from his or her ideas has grown into a considerable source of value, resulting in the creation of various mechanisms for protecting these property rights and in the study of IP management. The forms used for IP protection include nonlegal strategies such as lead-time, secrecy, the use of complementary sales and services, and statutory strategies such as copyrights, trademarks, industrial design rights, trade secrets (see, for example, Chally 2004), and patents.

Innovative companies view IP as both an asset and a challenge, where a new product or service may be the subject of a complex intersection of intellectual properties that includes the inventions that make it work, the trademarked branding that provides an identity, as well as the design, protected by copyright or design patent that gives it a recognizable look and feel. Protecting all of these intellectual properties and ensuring that a firm garners maximum value from its innovations is a difficult undertaking; thus the questions of whether and when to patent and how to avoid infringing the patents of others often occupy entire legal departments in large technology-based firms.

From a societal perspective, the state-sponsored monopoly created by a patent provides incentives for the development of innovations that may otherwise not emerge. However, it may result in inefficiencies that ultimately reduce social benefits if extended too broadly or for too long a period of time. Patenting has also been criticized where, for example, firms are sometimes seen as abusing the system for excessive gain by patenting preemptively in order to deter entry by others, and by removing technology that could provide society with common goods but would cannibalize their existing market (Agrawal and Garlappi 2007).

Firms often utilize a rigid one-size-fits-all approach to IP management, where all output from research and development (R&D) is protected (often through patents), without consideration of the specific context. Such a standardized approach creates inflexibility and thus entails significant costs, with the focus shifting from value created by the invention to cost recovery and legal protection. This may in turn inhibit external collaboration agreements and subvert open innovation practices, defined as "firms commercializing external (as well as internal) ideas by deploying outside (as well as in-house) pathways to the market" (Chesbrough 2003, 36–37). When there is limited expectation of cost recovery for creating the IP, or a deliberate attempt at creating barriers to the entry of competitors or increased switching costs for customers, inventions are not made available externally, thwarting attempts to build on such inventions and innovations. We therefore suggest that overly stringent IP policies or a prohibition on communication between internal and external researchers potentially discourages productive collaboration and the use of the IP by entrepreneurs, particularly when the invention targets a smaller market or is applicable in industrial sectors that lack resources necessary to manage IP.

Empirical studies on the value of patents have also yielded mixed results. For example, a study conducted for the National Bureau of Economic Research of R&D managers from 1,478 US manufacturing firms, supplemented by interviews with R&D managers and IP officers at nine firms, revealed that patents are perceived as more effective than other legal IP mechanisms, but significantly less effective than nonlegal mechanisms (Cohen, Nelson, and Walsh 2000). The study also found that patents offer more effective protection for product innovations than for process innovations, while secrecy is considered equally effective for both, especially for process innovations. Patents are the main mechanism for protecting product innovations in pharmaceuticals and medical equipment (50 percent of product innovations) and are considered important in special-purpose machinery, computers, auto parts, and miscellaneous chemicals (40 percent of product innovations), but they are held on only 27 percent of semiconductor technologies, 26 percent of communications equipment, and 21 percent of electronic components. The principal reasons for not patenting include the ease of legally inventing around a patent (cited by 65 percent of respondents), lack of novelty (55 percent), and patent filing information disclosure requirements (47 percent).

While it is widely acknowledged that industry sectors vary (Pavitt 1984; Alexy, Criscuolo, and Salter 2009), we argue that in most research conducted so far, IP management is treated in a relatively homogeneous manner – that is, a science-based model where strong institutions allow for rigid patent/legal protection with little ambiguity. However, the success of such an approach is dependent on the strength of the institutions, which vary across countries and in their ability to enforce such protection (North 1990), as discussed in the following section.

The Institutional Context of Intellectual Property Protection

As described above, intellectual property is understood to be both a private right and a public good. Patents (the most common form of IP protection today) are a social construction to encourage inventors to disclose and publicly share new knowledge in exchange for a limited-term monopoly, with the ultimate goal of diffusing useful inventions to achieve social benefits. It thus involves the interface between entrepreneurial incentives and how they are shaped by institutions, defined by Thomas Lawrence, Cynthia Hardy, and Nelson Phillips (2002, 282) as "relatively widely diffused practices, technologies, or rules that have become entrenched in the sense that it is costly to choose other practices, technologies, or rules."

Douglass North (1990) argues that economic growth is driven by the incentive structures that encourage individual effort and investment, which in turn are determined by institutions, that is, society's "rules of the game." William Baumol (1990) similarly suggests that entrepreneurial growth is heavily influenced by institutions, which in turn shape whether productive, unproductive, or destructive outcomes are encouraged. He defines innovation as the outcome of "productive entrepreneurship," where entrepreneurial activities lead to positive social gain, whereas "destructive entrepreneurship" is where entrepreneurial activities such as crime lead to more harm than good. Unproductive entrepreneurship is where entrepreneurs gain at the expense of others, through, for example, legal loopholes or rent seeking. Institutions should thus create an environment such that destructive and rent-seeking activities are discouraged and incentives for productive activities are encouraged. Thus, from an institutional perspective, patents can be seen as a mechanism that encourages productive entrepreneurship by balancing the entrepreneurial incentives with societal benefits, but are only as sound as the institutional setting in which they operate.

Leibenstein (1968, 78) suggests that "socio-cultural and political constraints" shape entrepreneurship within a given country, and such constraints include national, cultural, or universal values, relative wealth, type of government (for example, centralized planning versus open economy), population growth (Hunt and Levie 2003; Levie and Hunt 2004), and the economy's growth rate (Acs and Amoros 2008). Others see institutions as critical in reducing the uncertainty in a society as they encompass the structures that provide the incentives for different types of economic activity (DiMaggio and Powell 1983; Aldrich and Fiol 1994; North 1990). Acs and Szerb (2007) discuss the aspects of society-wide institutions implemented through public policy and regulation that are intended to be equally available to all people, but in practice access to the resources made available by such policy and regulation is limited for the poor and, more importantly, frequently these same institutional factors erect barriers to the poor. Two important specific examples of policy and regulations are property rights protection and access to the legal systems, which ultimately facilitate access to legitimate financial markets (Aidis 2005; Webb, Tihanyi, Ireland, and Sirmon 2009).

We suggest that the institutional environment shapes, and is shaped by, the level of formality of economic actors. Justin Webb and colleagues (2009) distinguish between those engaged in the formal economy (legal and legitimate means that produce legal and legitimate ends), the informal economy (producing legitimate ends but through illegal activities), and the renegade economy (illegal activities with illegitimate means and ends). They suggest that in environments where institutions are weak, alert entrepreneurs tend to find opportunities in the informal economy because the benefits of engaging in formal activities are low. North (1990) suggests that entrepreneurs weigh the incentives and restrictions in the environment as represented by regulations, or formal rules, as well as in terms of the prevailing cultural values and norms, or informal rules. Later, we will argue that poor farmers in developing countries face considerable challenges due to weak institutions, and as a result may be more inclined to engage in informal activities that are inconsistent with contemporary IP protection mechanisms (on this point, see also generally Chapter 4).

IP Management and Legitimization

In addition to the institutional environment, H. Aldrich and M. Fiol (1994) suggest that the acceptance of an innovation is dependent on its level of legitimacy, which they define in two dimensions. The first is "cognitive legitimacy," knowledge about the new activity and what is needed to succeed in an industry; the second is "socio-political legitimacy," the value placed on an activity by cultural norms and political influences. The former thus relates to the performance of the technology, whereas socio-political legitimization is influenced by the institutional environment in which it evolves.

Jeremy Hall and colleagues (2011) suggest that to fully understand the legitimacy dimensions of an innovation, the uncertainties under which the technology is being developed must also be fully understood. Those uncertainties have been categorized as either technological, commercial, organizational, or social (Freeman and Soete 1997; Hall and Martin 2005). Technological uncertainty has to do with overcoming scientific, technical, and engineering hurdles – that is, does the technology work? Commercial uncertainty has to do with whether the new technology can compete successfully in the marketplace – does it compete with less expensive or more effective alternatives? These two categories align with cognitive legitimacy, and, although often difficult, involve heuristics that are generally well understood.

Organizational uncertainty is concerned with whether the organization is able to appropriate the benefits of the technology – does the firm possess the required organizational capabilities, and is its IP strategy effective? Seminal work by David Teece (1986) suggests that even if a new product or process is technologically and commercially viable, innovators will not realize the benefits of the innovation unless they possess rigid IP protection and complementary assets, such as competitive manufacturing, complementary technologies, distribution, service, and other assets specific to the industry. He also suggests that strong complementary assets and efficacy of legal mechanisms of protection (that is, the appropriability regime) determine the likelihood that inventors will profit from their innovations. Organizations with a strong appropriability regime should attempt to commercialize the technology themselves, whereas those with weak regimes should license or sell the invention. More recently, David Teece, Gary Pisano, and Amy Shuen (1997) emphasized the dynamic nature of innovation, in that the value of innovation erodes into imitation or being replaced by newer innovations – what has been termed "Schumpeterian rents." They suggest that key to an innovative firm's survival and profitability is the ability to build, integrate, and reconfigure internal and external competencies to address rapidly changing environments.

Social uncertainty considers the impact on society of the technology and how a more diverse selection of secondary stakeholders may affect, or be affected by, such a development. As we shall show below, stakeholder concerns are playing an increasingly large role in new technology development. Jeremy Hall and Michael Martin (2005) suggest that social uncertainties differ from technological, commercial, and organizational uncertainties in that they involve more interacting variables with the inclusion of many stakeholders beyond the primary stakeholders identified in a firm's value chain. Furthermore, many of these social stakeholders are often difficult to identify, especially when technology has different meanings and interpretations within different social groups (Pinch and Bijker 1984), and particularly for secondary or tertiary stakeholders (Freeman 1984; Winner 1993), who may be geographically dispersed and whose identities are often unknown to the firm (Matos and Hall 2007; Hall and Vredenburg 2005). As a result, social uncertainties are more ambiguous in a situation where it is difficult to identify key variables and the concerns of potentially influential stakeholders are difficult to reconcile.

Hall and Martin (2005) suggest that these social uncertainties lead to higher complexity (more interacting variables) and ambiguity (inability to identify key variables), making it more difficult to determine the viability of the innovation. Understanding social uncertainty can be a particularly daunting task for scientifically and technologically oriented organizations that rely on scientific evidence-based heuristics that often do not align well with, or even consider, the ethical, religious, cultural, and/or social concerns of those that may be affected by the technology, as was the case with Monsanto (Shapiro 2000).

We posit that one solution to understanding the social concerns of stakeholders is to engage with them as early as possible during the technology's development. Such early knowledge will enable the developers to shape the technology to avoid controversy and better align with the needs of a broader range of users, thus improving legitimacy. Hall and colleagues (2011) suggest that overcoming technological and commercial uncertainties primarily involves cognitive legitimization processes, whereas social uncertainties are aligned with socio-political legitimization, the process by which key stakeholders accept a new venture, given their existing norms. They suggest that an innovation establishes legitimacy as its technical performance and social acceptance co-evolves and expands, thus reducing uncertainty. Drucker (2001) has suggested that in the future a key challenge for the corporation will be demonstrating its social legitimacy.

We also suggest that IP management policies play a crucial role in how legitimization evolves, but is a particularly challenging managerial task because it involves balancing both cognitive and socio-political legitimization processes. For example, from a cognitive perspective, increasing the use of patenting to block competition and provide bargaining chips for cross-licensing may provide strategic advantages for the firm but hinder the development of technologies that may result in longer-term social benefits, or result in the deferral of promising technologies, eroding socio-political legitimization. As we shall show below, IP management approaches that fail to consider socio-political legitimization processes are particularly vulnerable when the IP involves social controversy and weak institutions in, for example, developing and emerging economies. As a result, the use of IP management approaches based purely on cognitive legitimization may lead to a promising technology's being delayed or left sitting on the shelf.

Methodology

This study draws on primary and secondary data collected in Brazil, Canada, India, and the United States during the period 2003–12 as part of a broader research program investigating technological, commercial, organizational, and social issues related to innovation. Secondary data included desk research and information gathered from government documents and reports, the academic literature, and the international press, which enabled us to identify initial interview subjects. Other key stakeholders were identified by participants who suggested additional people to interview for the study – the snowball technique (Berg 1988). Primary data included interviews with a total of 192 subjects, including senior and middle managers, government officials, industry experts, officials of nongovernmental organizations, trade association members, academics, and other stakeholder representatives, such as 23 individual farmers and 4 farmer focus groups with 48 participants (see Table 6.1 for a detailed breakdown). A loosely structured style of face-to-face interviewing was applied, which enabled the research team to elicit key issues and perspectives on innovation management, IP approaches, public policy, stakeholder relations, and strategy. Interviews required between one and two hours, and were recorded and transcribed in summary form. A broad range of stakeholders with varying opinions regarding the technology were interviewed to help ensure data triangulation. Observation

TABLE 6.1

Interview subjects by stakeholder category

Stakeholder category	N
Industry representatives – senior company executives, middle managers, and trade association representatives	41
Farmers – individual interviews and four focus groups ·	71
Government officials – Brazil: Federal Ministries of Agriculture, Agricultural Development, and Science and Technology, National Technical Biosafety Commission (CTNBio), State Secretary of Agriculture, Brazilian Enterprise for Agricultural Research (EMBRAPA) officials, Brazilian Support Service to Micro and Small Enterprises (SEBRAE); Canada: Canadian Food Inspection Agency, Natural Resources Canada – Canadian Forest Service	37
United Nations officials – Food and Agriculture Organization (FAO); UN Environment Programme (UNEP); UN Economic Commission for Latin America and the Caribbean	9
NGOs and Farmer Associations– Athena Institute, Greenpeace, Sierra Club, Institute for Consumer Defense, Project Tamar, Council for Canadians, Polo Sindical da Borborema (Borborema Farmers Union), Esperança and Lagoa Seca (Brazil divisions)	11
University technology transfer offices – University of British Columbia, Industry Liaison Office	2
Community representatives	6
Academics	12
Consultants	3
Total	192

notes, manual coding of interviews, and archival data were used to best understand the interview results in the context of technological innovation.

We applied a case study approach using Monsanto's attempts to market transgenic technology in developing countries to illustrate how a rigid IP management approach may hinder the technology's legitimization process in complex environments (Yin 1994; Siggelkow 2007). This case is suitable because Monsanto is the main transgenic technology pioneer in agriculture and has marketed its technology in the developing regions in Brazil and the less developed areas of India. The case presentation first describes the business in nontechnical terms, followed by the contextual conditions of innovation in the agricultural biotechnology industry. We next discuss the organization's motivations, IP management policies, and challenges it encountered. Case analysis and discussions consider Monsanto's cognitive and socio-political legitimization processes and the role its IP strategies played in the diffusion of the technology.

The Challenge of IP Management in Agricultural Biotechnology

According to Joanna Chataway, Joyce Tait, and David Wield (2004), the adoption of biotechnology by the agrochemical industry was driven by competitive dynamics due to industry maturation as well as the need to address environmental issues. Indeed, agricultural biotechnology developers argued that the key feature of such an innovation was to reduce environmental impacts while enhancing agricultural output (Huang et al. 2002). Monsanto is widely regarded as the industry's biotechnology leader, employing 21,500 people in sixty-six countries, with sales of $13.5 billion and net income of $2.04 billion in 2012. However, it is also an example of a company that focused its IP management on establishing cognitive legitimacy, and as a result encountered considerable difficulties with its transgenic technology for agriculture in international markets.

Monsanto's Roundup Ready transgenic seeds (seeds with genes from other species) were developed to be resistant to the company's proprietary Roundup herbicides, which kill all plants except those bioengineered to resist the herbicide. Monsanto's value proposition for the farmer was based on a lower cost-to-output ratio, where farmers using such seeds require only one herbicide, allowing for greater efficiency (Magretta 1997). Even though the initial investment in the seeds and herbicide was more costly, the technology did not require any additional equipment or new skills (Hall et al. 2008, 47). Farmers would thus find Roundup Ready transgenic seeds relatively easy to adopt, as doing so would enable them to improve their business operations without eroding or rendering obsolete prior competencies (Hall and Martin 2005).

Managing organizational uncertainties was a high priority for Monsanto. For example, the company acquired the necessary competencies and complementary assets through acquisition of seed and biotech firms, and established alliances and networks to procure the needed IP rights (Hall and Crowther 1998). In addition, extensive legal mechanisms were established to effectively address patents, plant breeder's rights, trademarks, biological mechanisms such as hybridization, gene use–restriction technologies, and other means of preventing imitation or copying (Naseem,

Spielman, and Omamo 2010). The nature of the technology – namely, that the chemical makeup of the herbicides and genetic markers could easily be detected in the seeds – made it technologically feasible to use such a rigid IP management approach, where, for example, illegal use of Monsanto's products could be unequivocally determined. As a result, Monsanto was able to convince investors that the new technologies would radically change the industry and lead to promising returns in the long term (Chataway, Tait, and Wield 2004). Monsanto also convinced investors that its IP management approach and major investments in complementary assets would enable it to recoup its ongoing significant R&D investments, which totalled $1.52 billion in 2012.

Agribusinesses and large-scale farmers that were traditionally used to long-term contracts initially welcomed the technology (James 2003; Hall, Matos, and Langford 2007). Such organizations are accustomed to dealing with sophisticated contractual arrangements, and generally operate in countries with a strong institutional environment, such as the United States and Canada. Socio-political concerns related to the technology emerged in these countries, however. One example was a court dispute between Monsanto and Canadian canola farmer Percy Schmeiser on the rights of farmers to breed next-generation seeds versus Monsanto's rights to protect its IP. Even though the Supreme Court of Canada affirmed Monsanto's right to prevent farmers from using or breeding their transgenic seeds without permission (see *Monsanto Canada Inc. v. Schmeiser*, [2004] 1 S.C.R. 902, 2004 SCC 34), this case was often framed by the press as a David-and-Goliath battle between small farmers and powerful multinationals, reducing the social acceptability of the technology. Although a major controversy for Monsanto, the company's eventual victory reaffirmed as sound its IP management strategy based on cognitive legitimacy, albeit only within markets with strong institutions. Indeed, Monsanto's transgenic technology was widely adopted in North America, surpassing traditional varieties of soybeans, canola, corn, and other key commodity crops (Hall, Matos, and Langford 2007, 48).

More significant opposition to Monsanto's IP management policies emerged internationally, for example, with negative European public reaction to transgenics over concerns of vertically integrated multinationals that lacked public accountability (Tait 2001). The World Trade Organization's promotion of more liberal international trade was criticized by international public interest groups as supporting multinational companies at the expense of ordinary people and the environment, and was exploited by Monsanto and others as a way to force their crop products into Europe despite public opposition (Chataway, Tait, and Wield 2004; see also generally Chapter 1). Although these countries also possess strong institutions, such accusations emphasized the importance of socio-political issues, which complicated the technology's legitimacy.

Further complications in the adoption of the technology occurred in underdeveloped regions such as Brazil and India, where opponents claimed that foreign seed producers using rigid IP management policies could erode local farmer competencies, making them increasingly dependent on foreigners (Parayil 2003). Monsanto's struggles in Brazil and India can be seen as a form of "liability of foreignness," which has costs associated with spatial distance (travel, transportation, coordination, and so on), costs due to a firm's unfamiliarity with the local environment, home country restrictions on, for example, high-technology sales, "economic nationalism," and a lack of foreign firm legitimacy (Zaheer 1995). In Brazil, one of the world's major food exporters, soybeans were first considered unsuitable for tropical climates, but were then adapted for Brazil by the national research institute Brazilian Enterprise for Agricultural Research (EMBRAPA), with production costs also being lowered (Furtado 2004).[1] Brazil is now the world's second-largest soybean producer and exporter (after the United States), and the crop is expected to play a major role in the country's growing biodiesel sector (Hall et al. 2011).

Brazil has a long historical legacy of inequality (Griesse 2006). Small-scale farmers represent about 90 percent of the farms in Brazil, but occupy only 33 percent of the total agricultural area and produce only 40 percent of the gross agricultural production value.[2] Brazil's highly efficient large-scale agribusinesses are responsible for the bulk of the country's exports. This heterogeneity can significantly affect the impact institutions will have on entrepreneurship. For example, an impoverished entrepreneur living in circumstances where basic survival is the priority and informality is the norm will likely behave very differently from entrepreneurs from wealthier communities who have access to formal education, financial services, legal protection, and other institutional support. Thus, entrepreneurs from different income distributions will have different behaviour because they are exposed to different normative and cognitive institutions. Consequently, the introduction of Monsanto's transgenic soybeans into Brazil was highly controversial. Between 1998 and 2005, the country postponed the approval of the seeds, in part due to concerns that Monsanto's rigid IP approach would affect subsistence farming, whereas many large-scale farmers welcomed the technology and even illegally imported the seeds from Argentina (Hall, Matos, and Langford 2007).

According to our interview data, pressure from NGOs such as Greenpeace and the Brazilian-based Institute for Consumer Defense targeted the potentially detrimental impact of the technology on small farmers and national sovereignty, and the perception that Monsanto would ultimately control national resources. Indeed, the widespread distrust in Monsanto became a main deterrent to the technology's diffusion. As one senior EMBRAPA official stated, while senior managers perceived their joint venture with Monsanto as initially providing valuable complementary assets ("a match made in heaven"), EMBRAPA's reputation as a national technology contributor and champion was tarnished in the eyes of the Brazilian population, resulting in a "match made in hell." Inconsistent positions concerning the adoption of the technology were also seen at various government ministries. The Ministries of Agriculture (responsible for large-scale farming) and Science and Technology supported the technology, whereas the Ministries of Environment and Agricultural Development (responsible for small-scale, family, and subsistence farming) opposed it.[3]

In India, another important emerging economy for Monsanto's products, the company continues to deal with controversy over its attempts to protect proprietary seed traits. According to Herring (2006), opposition to Monsanto in India was based on concerns over threats to national independence, monopolization of seeds, potential "biological pollution," and potential undiscovered allergens that may affect human health. Monsanto was also blamed for encouraging high levels of farmer debt for its relatively more expensive transgenic *Bt* cotton, reportedly resulting in bankruptcies and suicides of Indian farmers (Shiva 2004) and exploitation of child labour (Venkateswarlu 2003).[4]

Overall, in contrast to Monsanto's domestic markets, which consist of large agribusinesses and relatively educated farmers familiar with contract law, proprietary issues were highly controversial for subsistence farmers in regions that lack experience with sophisticated contractual agreements. For example, in countries where institutions are strong, such as the United States and Canada, Monsanto forbids the testing or saving of seeds and effectively guards the intellectual property rights for its genetically modified products (as illustrated by the Schmeiser case, discussed above). Monsanto argues that the ability to save seeds, plant next-generation seeds, or conduct any independent testing falls outside farmers' rights as consumers (McKinney

2013). However, in countries where institutions are weak, such as India, Monsanto's seeds and herbicides are widely available without Monsanto's approval (Herring 2007; Chapter 4). Such practice is referred to as "stealth transgenics," where seeds "are saved, cross-bred, repackaged, sold, exchanged and planted in an anarchic agrarian capitalism that defies surveillance and control of firms and states," undermining international development concerns over biosafety and IP institutions (Herring 2007, 130).

Although Monsanto has the right to sue for patent infringement for such use, sale, and movement of "stealth seeds," legal action is not likely to be successful since most offenders are widely dispersed small-scale farmers, and the company's controversial status in India would engender little public support for such legal action. Although Monsanto's technology spread rapidly and widely in the country, the negative association with these controversies has eroded the socio-political legitimacy of the company and the technology, which in turn has affected the efficacy of any legal claims to its IP that Monsanto may wish to pursue.

Discussion

We have argued that following a rigid approach to IP management focused primarily on cognitive legitimization processes, particularly in international environments where institutions are weak, may result in the delayed diffusion of a promising technology, or the potential failure of a company's IP strategy. IP management should thus be viewed within a broader innovation management context, where an organization's policies toward IP can facilitate or hinder the overcoming of social uncertainties, which ultimately shape legitimization processes and influence technology diffusion. We investigated how different institutional settings affected Monsanto's international operations, and specifically how its IP management approach affected and was affected by social uncertainties in emerging and less developed economies.

Monsanto established itself as an industry technology leader through its extensive R&D investments, overcoming scientific hurdles and gaining cognitive legitimacy. It created an attractive business case for farmers in developed economies with strong institutional frameworks. For example, its products are now the industry standards in many mature market segments. Its rigid IP management strategies, based on, for example, enforceable patent protection, have enabled it to appropriate the benefits of its technology, although such strategies were controversial for some farmers in Monsanto's domestic markets.

Although Monsanto built the necessary organizational capabilities and complementary assets through its extensive experience in the agricultural sector, these capabilities and assets were insufficient for dealing with the social controversies that arose when it attempted to enter foreign markets, especially in Brazil and India. The company's inability to conform or adjust to cultural differences and to work within the informal rules commonly used by smaller farmers in poorer regions exacerbated the controversies and eroded the legitimacy of its technology. As a result, Monsanto has been unable to exploit institutional support internationally, and continues to struggle with social legitimacy, having acquired a reputation of being insensitive to the context of host countries and of attempting to control those countries' natural resources.

Monsanto's highly controversial business model, which is based on a rigid IP management strategy and focused mostly on establishing cognitive legitimacy, was not suitable for smaller operations in emerging and underdeveloped regions, resulting in costly delays and continued scrutiny regarding the societal benefits of its technology. Although Monsanto publicly discloses R&D investments of over $1 billion annually to establish cognitive legitimacy, the legal, administrative, and opportunity costs associated with the delays in Brazil and controversies in India (those related to sociopolitical legitimacy) remain less clear but are likely just as substantial.

The key to more effective technological innovation is early scanning of industry features and market dynamics, firm capabilities and appropriability issues, and potential social and environmental impacts (Hall and Martin 2005; Hall et al. 2011). Such scanning is consistent with the well-known "gatekeeper" concept of individuals who provide interfaces between the team staff and the outside environment (see generally Allen 1997). This includes technological gatekeepers who possess high levels of technical competence and keep informed by monitoring the scientific literature, attending conferences, and engaging in professional networking, and market gatekeepers who possess market knowledge and insights about how potentially exploitable market opportunities can be identified (O'Connor and McDermott 2004). We suggest that these gatekeeper roles need to be expanded to also address political and social awareness, and specifically how these roles are affected by institutional differences. Such capabilities may help firms identify external stakeholders who may fear potential negative impacts of a new technology, and address their concerns at an early phase of the technology's development or even identify socially beneficial applications that will enhance the technology's legitimacy. However, further research is needed to determine whether it is possible to simultaneously apply such early scanning approaches under rigid IP policies. We speculate that more open innovation approaches to IP management may be an alternative for providing technology developers with insights from various stakeholders at earlier stages of development, improving legitimization, and potentially reducing controversy, although we also acknowledge that such approaches may not overcome the financial constraints necessary for such ambitious technology.

Moving a firm more toward open innovation is about finding and exploiting expertise and competencies wherever they may be, and then sharing the knowledge and information as freely as possible. Companies can employ techniques, from crowd sourcing through Internet forums that encourage problem solving from anyone in the world (Huston and Sakkab 2006; Judson and Dapprich 2009), to integrated outsourcing, to key employee secondment, and these open innovation methods can be applied at many points in the value chain. However, most literature on crowd sourcing is focused on cognitive legitimization processes, such as overcoming technological and commercial challenges. We therefore suggest that further research is needed to understand how the open innovation concept can be used to incorporate a broader stakeholder perspective as early in the development cycle as possible.

Embracing open innovation does not necessarily preclude the use of the more traditional methods of partnerships, mergers, and acquisitions for reaching outside the firm while retaining control and intellectual property protection. Nor does it necessarily replace all forms of IP. Indeed, it is unlikely that Monsanto would have been able to initially generate capital for its investments without a relatively rigid IP strategy. It is important to find the balance that works best for the specific context of the innovation, as open innovation does not work the same way in all industries but rather needs to be aligned with the overall business model of the firm (Seldon 2011). In biotech, Paul Nakagaki, Josh Aber, and Terry Fetterhoff (2011, 33) reported that at Roche "we identified two important elements that affect our company's ability to embrace open innovation: creating the compelling eureka moment that will inspire senior management to champion open innovation and changing the mindset to create an open innovation culture that pervades the organization." We suggest that rigid approaches to IP management focused primarily on cognitive legitimization may result in the delayed diffusion of promising technology, or even the potential failure of a firm's IP strategy. Indeed, the multiplicity of options available to innovative firms to protect the fruits of their labours and overcome hurdles to dissemination, both formal and informal, exemplify aspects of the IP–Regulatory Complex.

As demonstrated in this chapter, where key formal institutions – including government regulations, oversight, and enforcement – are weak, the aforementioned delays and failures may be felt more keenly. In the context of Monsanto's IP management strategies, we suggest that their inability to address the weakness of formal institutions, to conform or adjust to cultural differences, and to work within the informal rules commonly used by smaller farmers in poorer regions ultimately exacerbated controversy and contributed to eroding the legitimacy of the technology. As discussed throughout this volume, early consideration of the collective impact of regulatory regimes and intellectual property protection mechanisms that may be relevant to the development and dissemination of a given technology may assist firms in developing an overall IP management strategy that is appropriate to that technology in all the circumstances. This should include an accounting of informal institutions and relevant cultural and societal considerations.

This study presents a critique on the complex challenges of managing IP for agricultural biotechnology in international and often controversial agricultural sectors. While the Monsanto example used here is not the only situation where IP can be challenging, it is an illustration of the types of challenges that innovative firms face when addressing IP protection and its relationship to the legitimization of innovation.

Monsanto's rigid IP management strategy, while effective for cognitive legitimization, alienated local farmers in developing regions, eroding the overall legitimacy of the technology. Balancing cognitive and socio-political legitimization processes is thus necessary for IP management. We suggest this can be done by engaging with those that may be affected by the technology, through, for example, expanding the scope of technological and market gatekeeper roles to include political and social awareness.

Notes

1 EMBRAPA = Empresa Brasileira de Pesquisa Agropecuária (in Portuguese).
2 Interview, Ministry of Agricultural Development, November 25, 2003.
3 Personal communication, Coordinator, Genetic Resources Conservation and Sustainable Use, Ministry of Environment; and Assistant to the Minister of Agricultural Development, Brasilia, 2003. See also Hall, Matos, and Langford 2008.
4 *Bt* cotton is transgenic cotton that contains the bacterium *Bacillus thuringiensis*, which produces a chemical harmful to various pests.

References

Secondary Sources

Acs, Zoltan J., and J.E. Amoros. 2008. "Entrepreneurship and Competitiveness Dynamics in Latin America." *Small Business Economics* 31 (3): 305–22. http://dx.doi.org/10.1007/s11187-008-9133-y.

Acs, Zoltan J., and Laszlo Szerb. 2007. "Entrepreneurship, Economic Growth and Public Policy." *Small Business Economics* 28 (2-3): 109–22. http://dx.doi.org/10.1007/s11187-006-9012-3.

Agrawal, Ajay K., and Lorenzo Garlappi. 2007. "Public Sector Science and the Strategy of the Commons." *Economics of Innovation and New Technology* 16 (7): 517–39. http://dx.doi.org/10.1080/10438590600914627.

Aidis, Ruta. 2005. "Institutional Barriers to Small- and Medium-Sized Enterprise Operations in Transition Countries." *Small Business Economics* 25 (4): 305–17. http://dx.doi.org/10.1007/s11187-003-6463-7.

Aldrich, H., and M. Fiol. 1994. "Fools Rush In? The Institutional Context of Industry Creation." *Academy of Management Review* 19 (4): 645–70.

Alexy, Oliver, Paola Criscuolo, and Ammon Salter. 2009. "Does IP Strategy Have to Cripple Open Innovation?" *MIT Sloan Management Review* 51 (1): 71–77.

Allen, T. 1997. *Managing the Flow of Technology.* Cambridge, MA: MIT Press.

Baumol, William J. 1990. "Entrepreneurship: Productive, Unproductive, and Destructive." *Journal of Political Economy* 98 (5): 893–921. http://dx.doi.org/10.1086/261712.

Berg, S. 1988. "Snowball Sampling." In *Encyclopedia of Statistical Sciences,* 8th ed., ed. S. Kotz and N.L. Johnson. New York: Wiley.

Chally, Jon. 2004. "The Law of Trade Secrets: Toward a More Efficient Approach." *Vanderbilt Law Review* 57 (4): 1269–1311.

Chataway, Joanna, Joyce Tait, and David Wield. 2004. "Understanding Company R&D Strategies in Agro-Biotechnology: Trajectories and Blind Spots." *Research Policy* 33 (6–7): 1041–57. http://dx.doi.org/10.1016/j.respol.2004.04.004.

Chesbrough, Henry W. 2003. "The Era of Open Innovation." *MIT Sloan Management Review* 44 (3): 35–41.

Cohen, W.M., R.R. Nelson, and J.P. Walsh. 2000. *Protecting Their Intellectual Assets: Appropriability Conditions and Why US Manufacturing Firms Patent (or Not).* Cambridge, MA: National Bureau of Economic Research. http://dx.doi.org/10.3386/w7552.

DiMaggio, P., and W. Powell. 1983. "The Iron Cage Revisited: Institutional Isomorphism and Collective Rationality in Organizational Fields." *American Sociological Review* 48 (2): 147–60. http://dx.doi.org/10.2307/2095101.

Drucker, P. 2001. "The Next Society." *Economist* 361 (8246): 3–5.

Freeman, C., and L. Soete. 1997. *The Economics of Industrial Innovation.* London: Pinter.

Freeman, R. 1984. *Strategic Management: A Stakeholder Approach.* Boston: Pitman.

Furtado, R. 2004. "A pesquisa que revolucionou a agricultura." *Scientific American Brasil* 22 (March). http://www2.uol.com.br/sciam/reportagens/a_pesquisa_que_revolucionou_a_agricultura.html.

Griesse, M. 2006. "The Geographic, Political, and Economic Context for Corporate Social Responsibility in Brazil." *Journal of Business Ethics* 73: 21–37.

Hall, Jeremy, and Sarah Crowther. 1998. "Biotechnology: The Ultimate Cleaner Production Technology?" *Journal of Cleaner Production* 6 (3–4): 313–22. http://dx.doi.org/10.1016/S0959-6526(98)00006-7.

Hall, Jeremy, and Michael Martin. 2005. "Disruptive Technologies, Stakeholders and the Innovation Value-Added Chain: A Framework for Evaluating Radical Technology Development." *R&D Management* 35 (3): 273–84. http://dx.doi.org/10.1111/j.1467-9310.2005.00389.x.

Hall, Jeremy, Stelvia Matos, and Cooper Langford. 2007. "Social Exclusion and Transgenic Technology: The Case of Brazilian Agriculture." *Journal of Business Ethics* 77 (1): 45–63. http://dx.doi.org/10.1007/s10551-006-9293-0.

Hall, Jeremy, Stelvia Matos, Bruno Silvestre, et al. 2011. "Managing Technological and Social Uncertainties of Innovation: The Evolution of Brazilian Energy and Agriculture." *Technological Forecasting and Social Change* 78 (7): 1147–57. http://dx.doi.org/10.1016/j.techfore.2011.02.005.

Hall, Jeremy, and Harrie Vredenburg. 2005. "Managing Stakeholder Ambiguity." *MIT Sloan Management Review* 47 (1): 11–13.

Hanel, Petr. 2006. "Intellectual Property Rights Business Management Practices: A Survey of the Literature." *Technovation* 26 (8): 895–931. http://dx.doi.org/10.1016/j.technovation.2005.12.001.

Herring, Ronald J. 2006. "Why Did 'Operation Cremate Monsanto' Fail? Science and Class in India's Great Terminator-Technology Hoax." *Critical Asian Studies* 38 (4): 467–93. http://dx.doi.org/10.1080/14672710601073010.

–. 2007. "Stealth Seeds: Bioproperty, Biosafety, Biopolitics." *Journal of Development Studies* 43 (1): 130–57. http://dx.doi.org/10.1080/00220380601055601.

Huang, J., R. Hu, S. Rozelle, et al. 2002. "Transgenic Varieties and Productivity of Smallholder Cotton Farmers in China." *Australian Journal of Agricultural and Resource Economics* 46 (3): 367–87. http://dx.doi.org/10.1111/1467-8489.00184.

Hunt, Stephen, and Jonathan Levie. 2003. "Culture as a Predictor of Entrepreneurial Activity." In *Frontiers of Entrepreneurship Research: Proceedings of the Twenty-Third Annual Entrepreneurship Research Conference*, 171–85. Babson Park, FL: Babson College.

Huston, L., and N. Sakkab. 2006. "Connect and Develop: Inside Procter & Gamble's New Model for Innovation." *Harvard Business Review* 84 (3): 58–66.

James, Harvey S. Jr. 2003. "On Finding Solutions to Ethical Problems in Agriculture." *Journal of Agricultural and Environmental Ethics* 16 (5): 439–57. http://dx.doi.org/10.1023/A:1026371324639.

Judson, E., and J. Dapprich. 2009. "Negotiation 2.0." *Nature Biotechnology* 27: 1–3.

Lawrence, Thomas B., Cynthia Hardy, and Nelson Phillips. 2002. "Institutional Effects of Interorganizational Collaboration: The Emergence of Proto-Institutions." *Academy of Management Journal* 45 (1): 281–90. http://dx.doi.org/10.2307/3069297.

Leibenstein, Harv. 1968. "Entrepreneurship and Development." *American Economic Review* 58 (2): 72–83.

Levie, Jonathan, and Stephen Hunt. 2004. "Culture, Institutions, and New Business Activity: Evidence from Global Entrepreneurship Monitor." In *Frontiers of Entrepreneurship Research 2004: Proceedings of the Twenty-Fourth Annual Entrepreneurship Research Conference*, 519–33. Babson Park, FL: Babson College.

Magretta, Joan. 1997. "Growth through Global Sustainability." *Harvard Business Review* 75 (1): 78–88.

Matos, Stelvia, and Jeremy Hall. 2007. "Integrating Sustainable Development in the Supply Chain: The Case of Life Cycle Assessment in Oil and Gas and Agricultural Biotechnology." *Journal of Operations Management* 25 (6): 1083–1102. http://dx.doi.org/10.1016/j.jom.2007.01.013.

McKinney, K. 2013. "Troubling Notions of Farmer Choice: Hybrid Bt Cotton Seed Production in Western India." *Journal of Peasant Studies* 40 (2): 351–78. http://dx.doi.org/10.1080/03066150.2012.709847.

Nakagaki, P., J. Aber, and T. Fetterhoff. 2012. "The Challenges in Implementing Open Innovation in a Global Innovation-Driven Corporation." *Research Technology Management* 55 (4): 32–38. http://dx.doi.org/10.5437/08956308X5504079.

Naseem, Anwar, David J. Spielman, and Steven Were Omamo. 2010. "Private-Sector Investment in R&D: A Review of Policy Options to Promote its Growth in Developing-Country Agriculture." *Agribusiness* 26 (1): 143–73. http://dx.doi.org/10.1002/agr.20221.

North, Douglass C. 1990. *Institutions, Institutional Change and Economic Performance.* Cambridge: Cambridge University Press. http://dx.doi.org/10.1017/CBO9780511808678.

O'Connor, G.C., and C.M. McDermott. 2004. "The Human Side of Radical Innovation. *Journal of Engineering and Technology Management* 21 (1–2): 11–30.

Parayil, Govindan. 2003. "Mapping Technological Trajectories of the Green Revolution and the Gene Revolution from Modernization to Globalization." *Research Policy* 32 (6): 971–90. http://dx.doi.org/10.1016/S0048-7333(02)00106-3.

Pavitt, Keith. 1984. "Sectoral Patterns of Technical Change: Towards a Taxonomy and a Theory." *Research Policy* 13 (6): 343–73. http://dx.doi.org/10.1016/0048-7333(84)90018-0.

Pinch, Trevor J., and Wiebe E. Bijker. 1984. "The Social Construction of Facts and Artifacts: Or How the Sociology of Science and the Sociology of Technology Might Benefit Each Other." *Social Studies of Science* 14 (3): 399–441. http://dx.doi.org/10.1177/030631284014003004.

Seldon, T. 2011. "'Beyond Patents: Effective Intellectual Property Strategy in Biotechnology." *Policy and Practice* 13 (1): 55–61. http://dx.doi.org/10.5172/impp.2011.13.1.55.

Shapiro, Robert. 2000. "The Welcome Tension of Technology: The Need for Dialogue about Agricultural Biotechnology." St. Louis: Center for the Study of American Business, Washington University.

Shiva, Vandana. 2004. "The Suicide Economy of Corporate Globalisation," Counter Currents, http://www.countercurrents.org/glo-shiva050404.htm.

Siggelkow, N. 2007. "Persuasion with Case Studies." *Academy of Management Journal* 50 (1): 20–24. http://dx.doi.org/10.5465/AMJ.2007.24160882.

Tait, Joyce. 2001. "More Faust than Frankenstein: The European Debate about the Precautionary Principle and Risk Regulation for Genetically Modified Crops." *Journal of Risk Research* 4 (2): 175–89. http://dx.doi.org/10.1080/136698700 10027640.

Teece, David J. 1986. "Profiting from Technological Innovation: Implications for Integration, Collaboration, Licensing and Public Policy." *Research Policy* 15 (6): 285–305. http://dx.doi.org/10.1016/0048-7333(86)90027-2.

Teece, D.J., G. Pisano, and A. Shuen. 1997. "Dynamic Capabilities and Strategic Management." *Strategic Management Journal* 18 (7): 509–33. http://dx.doi.org/ 10.1002/(SICI)1097-0266(199708)18:7<509::AID-SMJ882>3.0.CO;2-Z.

Venkateswarlu, Davuluri. 2003. "Child Labour and Trans-National Seed Companies in Hybrid Cotton Seed Production in Andhra Pradesh." India Committee of the Netherlands (ICN), http://www.indianet.nl/cotseed.html.

Webb, Justin W., Laszlo Tihanyi, R. Duane Ireland, and David G. Sirmon. 2009. "You Say Illegal, I Say Legitimate: Entrepreneurship in the Informal Economy." *Academy of Management Review* 34 (3): 492–510. http://dx.doi.org/10.5465/AMR. 2009.40632826.

Winner, Langdon. 1993. "Upon Opening the Black Box and Finding It Empty: Social Constructivism and the Philosophy of Technology." *Science, Technology and Human Values* 18 (3): 362–78. http://dx.doi.org/10.1177/016224399301800306.

Yin, R. 1994. *Case Study Research: Design and Methods.* 2nd ed. Beverly Hills, CA: Sage Publications.

Zaheer, Srilata. 1995. "Overcoming the Liability of Foreignness." *Academy of Management Journal* 38 (2): 341–63. http://dx.doi.org/10.2307/256683.

PART 3

FUTURE DIRECTIONS

7

A Governance Approach to the Agricultural Genomics Intellectual Property–Regulatory Complex

REGIANE GARCIA

The objective of this chapter is to stimulate discussion on how a new governance model that includes participation of nonstate actors in agricultural genomics policy making can contribute to the development of more responsive policies.

The Intellectual Property–Regulatory Complex faces countless challenges in order to facilitate the delivery of agricultural genomics innovations to end-users. Secrecy and exclusion of affected stakeholders in the design and operationalization of the IP–Regulatory Complex are two such challenges (see, for example, Chapters 2, 4, and 8). Secrecy and exclusion of a plurality of viewpoints and interests during the creation of a regulatory framework or during policy making can result in important considerations that are relevant to many stakeholders being overlooked. For example, the development of Canada's agricultural biotechnology framework in the 1990s overlooked the participation of the public and marginalized important social and ethical concerns, as Sarah Hartley discusses in Chapter 2.

One of the tasks faced by the contributors to this book was to propose solutions to "facilitate greater efficiency in meeting the aims of IP and regulatory regimes while enabling innovative research" to reach end-users. Some of the earlier chapters have proposed solutions to problems that result from the complexities of the multilayered legal landscape governing agricultural genomics.

This chapter does not propose ways of handling such problems. Instead, it takes a step back and focuses on the problems related to the institutional arrangements that create the complexities of the legal framework, particularly the lack of stakeholders' participation in the design of policies governing agricultural innovations in Canada. It proposes that one way to deal with the effects of stakeholders' exclusion is to create more democratic and inclusive policy-making processes and arenas. It suggests that a paradigmatic shift from a regulatory model to a "new governance model" based on citizens' and stakeholders' inclusion and participatory governance are more likely to respond to the needs of different end-users in a timely and context-sensitive fashion in Canada.

This chapter begins with an overview of the literature discussing the paradigm shift from regulation, based on centralized decision making by experts, to new governance, grounded in multistakeholder involvement in policy making and management. The overview includes the origins of each model and the problems new governance attempts to address. This is followed by a discussion of the relationship between a new governance model and the traditional regulatory approach, including a brief discussion of the interaction of new governance and the traditional regulatory model. Drawing on the literature of governance design and on an ongoing experiment that mixes the new and old models of governance, some examples are offered of ongoing governance experiments that might be applicable to Canada's agricultural genomics arena. In conclusion, we are reminded that major changes in the ways in which governments decide on and carry out public goals are on the way, and that proposals to facilitate greater efficiency in meeting the aims of the IP–Regulatory Complex should take into account these fundamental challenges.

It is important to keep in mind that the purpose of this chapter is simply to invite the reader to think of a different way to address the conflicting world views and interests embedded in the agricultural genomics legal arena. It is also important to keep in mind that the idea of "new governance" presented here has yet to be tested in the specific arena of genomics, and, as with all novelties, there are many open questions as to how and whether such ideas might work effectively in practice. Please note, however, that new governance is not presented either as the solution to all challenges that impede the IP–Regulatory Complex or as a substitute for the regulatory model. In fact, the chapter acknowledges some of the shortcomings and difficulties concerning the operationalization of new governance frameworks and of some models in which the regulatory model is combined with the new governance model. The ideas in this chapter emerge from my own research on human rights, public health policy making, and health system governance, in which I have had to continuously deal with different and often conflicting rules, world views, and interests that often impede collaboration on designing the best solutions for all stakeholders, despite people's best efforts and intentions. My hope is that this chapter leads to an in-depth inquiry into the significance and effectiveness of a new governance approach to the agricultural genomics landscape, while bringing to bear insights from other areas of knowledge and experience as a means of facilitating the ultimate goal of delivering innovative genomics research to end-users in Canada.

From Regulation to New Governance

The regulatory model originated in the New Deal programs of the 1930s. The New Deal was a series of domestic legislative and administrative programs in the United States, either passed by Congress or enacted by presidential executive order, with the goal of responding to the economic downturn of the Great Depression. Many administrative agencies were established to work with the federal government to ensure that the programs were implemented and regulated. It is important to note, however, that an in-depth discussion of the New Deal and its programs is not relevant for the purpose of this chapter. The intent is only to show that the new governance approach emerged as a response to the administration model of the New Deal era.

The objective of the New Deal (or regulatory model) was to promote scientific rationality, efficiency, homogeneity, uniformity, and central coordination of the New Deal programs (Lobel 2004). In the regulatory model, the legislature is responsible for enacting laws but administrative agencies made up of experts are responsible for policy decision making to give effect to those laws. The belief was that the advancement of scientific knowledge and technological inventions occurs much more rapidly than the process of political decision making, and that only scientific experts could comprehend complex policy issues. In the regulatory model, experts are the dominant actors, and there is no need for public participation in the production of knowledge and policies (Fischer 2008). In fact, advocates of the regulatory model advise against including laypersons in policy making, claiming that their involvement would bring about quarrels and therefore be ineffective and inefficient (Hart 2003). The institutional design of policy making should ensure that only "elite scientists," who understand the "best available knowledge," are involved in the identification and design of the best way to solve problems (Fischer 2008).

Since the 1970s, however, this expert-regulatory model has been increasingly criticized for its lack of transparency and inattention to a plurality of viewpoints and interests. New approaches to governance with public involvement in policy making have emerged as alternatives for handling problems that the traditional regulatory model is incapable of handling (Hage, Leroy, and Willems 2006).[1] In the United States, for instance, critics of expert governance have called for alternatives to regulation and for inclusion of the public in the policy-making arena (Tushnet 2003). As Frank Fischer (1993) points out, citizens took to the streets to protest centralized and technocratic expert policy decisions. Fischer (1993, 295) notes that

> such demonstrations have occurred against scientists performing chemical tests on unknowing citizens, risk assessors pronouncing the safety of a hazardous waste incinerator in a highly populated neighborhood, doctors performing (or not performing) abortions, regulators withholding experimental treatments for AIDS, biotechnologists genetically altering and irradiating foods, medical researchers performing unnecessary experiments on animals, and physicians denying their patients' requests to die, educators busing children to distant neighborhoods, psychologists and social workers telling families how to raise their kids, among others.

The dissatisfaction with centralized and technocratic policy-making processes was not restricted to the United States; many Latin American countries also rallied against the expert, state-centred, top-down, and one-size-fits-all model of policy making (Richardson, Sherman, and Gismondi 1993). In Brazil, for example, which was under military dictatorship at the time, open protests against technocratic and centralized policy making called not only for a return to democracy but also for more citizen participation in public policy making and public management (Costa 1986).

In fact, across the globe, citizens and social movements in the 1970s expressed their concerns not only with technocratic and centralized political decision making but also with the degree to which scientists had vested interests in their own research projects and exhibited a disregard for the possible consequences of their work (Fischer 1993). At the same time, "members of the public [and some scientists themselves] increasingly [came] to see that scientists are themselves laypersons in matters of political goals and social judgments, and increasing numbers speak of the need for democratic regulation and control of science" (Fischer 1993, 296–97). As Habermas elaborates, policy problems cannot be solved by scientific evidence alone; rather, they require a critical interdependence and permanent communication between experts, politicians, and the public (Habermas 1970). Furthermore, modern societies are full of expert dissent, scientific uncertainty, and scientific ignorance (Fischer 2008). Silvio Funtowicz and Jerome Ravetz (1993, 379) term uncertainty and ignorance as "post-normal science," which they define as "based on assumptions of unpredictability, incomplete control, and plurality of legitimate perspectives ... where facts are uncertain, values are in dispute, stakes are high and decisions urgent." Accordingly, neither experts nor politicians should be the only actors in policy-making arenas; rather, the public should participate in political decisions as well.

It was in this context of dissatisfaction with scientific professional expertise that calls emerged for more collaborative and participatory research in academic settings (Edelstein 1988), direct citizen participation in political decision making (Arnstein 1969), and the development of more democratic political systems (Habermas 1970). Furthermore, developments in technology and communications, changes in economy, polity, and society, and the rise of nonstate actors in national and supranational arenas have contributed to the move from a regulatory to a new governance model of policy making (Lobel 2004, 266–67). This policy-making shift became even more common in policy fields with multifaceted and intertwined problems, such as policies concerning the environment and ethics in health services (Fung and Wright 2001). Instead of expert state actors in regulatory agencies developing centralized, universal, and coercive policies, a multitude of nonstate actors became involved in the process of policy making as well, bringing their own perspectives and insights to problems and participating in the design of the best solutions to these problems (Fung and Wright 2001).

This new way of policy making and approach to problem solving is referred to as "governance," which can be described as "a gradual transformation of the scientific-political domain where the formulation of goals, the choice of instruments and the implementation of solutions becomes a combined task of political actors and institutions, corporate interests, civil society and transnational organizations" (Hage, Leroy, and Willems 2006, 6). These important features of new governance – a shift in policy-making processes and the reliance on guidelines, benchmarks, and standards without formal sanctions – have raised questions about the interactions between the traditional and new models of governance, which we deal with next.

Interaction between the New and the Traditional
Governance Models

The understanding of the relationship between the new and traditional governance models is in its early stages (Trubek and Trubek 2010), but it is possible to identify a conviction among legal scholars that governance requires one to think of law and its nature beyond core concepts such as a state-centred, top-down, one-size-fits-all enterprise (Trubek and Trubek 2010). More specifically, David and Louise Trubek write (2010, 721):

> [Legal scholars] sought to characterize new governance as much by the way it differed from some image of law as by a positive definition. Efforts to describe the new phenomenon might suggest that it is an approach that is normative but not legally binding; it is designed to form norms through bottom up processes not by top- down legislation; it uses open-ended rather than precise legal rules; and it is flexible and revisable rather than fixed.

For a clearer understanding, it is important to underline some legal features of a new governance approach that are relevant to the present discussion. In previous work, Louise Trubek (2006, 11) describes the new governance phenomenon as a process "designed to increase flexibility, improve participation, [and] foster experimentation and deliberation." In order to achieve these goals, a new governance model requires new institutional arrangements, such as a devolution of decision-making power from central to local bodies, more flexible rules, the creation of stakeholder networks for coordination, and increased dissemination of information among the networks (Salamon 2001). Trubek (2006, 12) further elaborates that: (1) devolution enables local governments to make context-sensitive decisions; (2) standards and benchmarks allow experimentation with problem-solving initiatives and appropriate adjustment as new information is produced; and (3) stakeholder networks foster collection, coordination, and dissemination of information and function as vehicles for stakeholders to voice different positions and perspectives. With these new institutional arrangements, new political actors, and reliance on guidelines and benchmarks without formal sanctions ("soft law"), new governance is said to be transformative of law insofar as these new institutional arrangements challenge what we think of as law (Trubek 2006, 13).

Orly Lobel (2004), more specifically, describes ways in which law and lawmaking processes have changed under the new governance model. The nature of law in the traditional regulatory model falls into a general category of law referred to as "command-and-control," whereby state actors (the legislature or executive and regulatory agencies) are authorized to regulate an activity by enacting statutes or regulations that prescribe what is permitted and what is illegal (Lobel 2004). The "command" part refers to the detailed prescription of standards that one must comply with, whereas the "control" part refers to the adverse sanctions that one may experience in the event of noncompliance (Baldwin, Cave, and Lodge 2012). This traditional model of regulation involves detailed legal prescription in a centralized, top-down, and rigid structure, and develops uniform and generalized rules (Lobel 2004, 325). Noncompliance with, or violation of, a legal provision may be subject to penalties. However, in a new governance model that allows more flexible rules based on guidelines, benchmarks, and standards, the nature of law becomes procedural, reflexive, decentralized, flexible, adaptable, and contextualized (Lobel 2004, 325). In governance settings, therefore, stakeholders can more freely "experiment with solutions" and make adjustments if necessary without the threat of legal consequences. Nevertheless, despite a seemingly shared understanding that new governance defies what we traditionally think of as law, the exact nature of the law and the interaction of traditional and new governance mechanisms remain contentious and not yet fully developed.[2]

In a preliminary mapping exercise of existing configurations between the old and new forms of governance, David and Louise Trubek (2007, 6) identify three main arrangements: complementarity, rivalry, and transformation or hybridity. In the complementarity arrangement, regulatory and new governance systems work together for common goals; in the rivalry arrangement, the regulatory and new governance systems compete for dominance; and in the transformation or hybridity arrangement, Trubek and Trubek depict the regulatory and new governance systems as being merged into a new single process. As they cautiously remark, however, the state of the art in the context of legal scholarship is still unclear (Trubek and Trubek 2007, 3):

> There is a vigorous debate about new governance. Some [scholars] think that these innovations should be used only in a very limited way to supplement traditional forms of regulation in areas in which command and control processes have not been effective. Others think that we are witnessing a major transformation in law and policy, and that the new governance "revolution" will end up changing all law as we know it.

Despite a lack of understanding about the interactions between regulation and new governance, it is impossible and counterproductive to overlook the new ways in which governments carry out public goals directed at complex political problems. As noted above, the shift from regulation to governance occurred largely because political institutions were unable to satisfactorily address issues of modern social complexity (Kooiman 2003). The problem is that "the institutional forms of liberal democracy developed in the nineteenth century – representative democracy plus techno-bureaucratic administration – seem increasingly ill-suited to the novel problems we face in the twenty-first century," as Archon Fung and Erik Olin Wright (2001, 5) put it. Many contributors to this book appear to agree that in the context of agricultural genomics in Canada and internationally, the secrecy of the law and policy making and the lack of stakeholders' inclusion in the decision-making processes have been problematic (see, for example, Chapters 2, 4, and 8). A vast literature proposes that a "more participatory" concept of democracy grounded on citizens' participation in policy making (Barber 1984) works better to deal with issues related to the modern social complexity of our days, including diverse world views, and the fact that there is not a set of directions for designing one-size-fits-all regulation for complex and multifaceted problems (Ayres and Braithwaite 1992). The agricultural genomics arena is a case in point that involves many stakeholders with different world views and diverse frames for understanding the issues, which makes it hard to design a set of directions that serves all interests.

The remainder of this chapter underlines skeptical arguments against new governance and suggests the *European Union Water Framework Directive* as a new governance experiment that can provide some insights on new governance institutional arrangements for Canada's agricultural genomics legal arena.

New Governance in Practice

A new governance approach with broad stakeholder participation in policy making is more suitable for addressing agricultural genomics issues than the existing regulatory model. Participatory policy making includes the public and helps create a forum for exchange of diverse world views between citizens, politicians, and experts, which is important to handling social complexity and opinions concerning agricultural genomics innovation.

Skepticism and Proposals to Mitigate Skeptics' Concerns

Such a model has been criticized, however, for its cheerful belief in civil society, for a lack of precision concerning what the public can add to scientific discussion, and for derailing the process with irrational and emotional discussions (Fischer 2008). Along the same lines, James Johnson (1998) suggests that it is unreasonable to assume that deliberation will guarantee reasonable arguments and actions or good faith on the part of participants. Others have pointed out the challenges of genuinely including marginalized populations in the process. Lynn Sanders (1997) notes, for example, that the deliberation project is said to enhance autonomy; educate people to consider broader political questions; avoid expert driven policy making; include everyone; and potentially improve individuals as human beings, citizens (as political actors), and members of a community (in pursuit of a common voice). She stresses, however, that despite these commendable goals, it is difficult in reality to ensure that excluded groups, particularly disenfranchised ones, will have a genuine opportunity to influence the decision-making process, largely due to imbalances of knowledge and power in deliberative forums (Sanders 1997).

Institutional Design Matters

To ensure the meaningful inclusion of ordinary people in the solution of problems, theorists and practitioners of deliberative democracy and conflict resolution argue that adequate institutional design is remarkably important. Building on the work of Jürgen Habermas, Archon Fung and Erik Wright have developed guidelines for institutional arrangements to improve democratic decision making. In "Deepening Democracy: Innovations in Empowered Participatory Governance," Fung and Wright (2001) examine five such experiments. The common features among the experiments were a commitment to practices and values of communication, public justification, and deliberation; a commitment to civil society participation; and institutional vehicles of public participation directly associated with government bodies (e.g., advisory bodies with decision-making power). In their words, "although each of these experiments differs from the others in its ambition, scope, and concrete aims, they all share surprising similarities in their motivating principles and institutional design features" (Fung and Wright 2001, 17). The motivating principles are: (1) a focus on specific problems; (2) involvement of ordinary people and officials; and (3) deliberative decision making. The institutional arrangement features are: (1) devolution of decision-making authority; (2) creation of formal linkages of coordination with more centralized authorities; and (3) creation of institutional participation mechanisms (Fung and Wright 2001). Applying the institutional structures and animating principles above, all five experiments successfully included the most affected populations, including ordinary citizens and officials in the field, to offer a variety of experience and knowledge and collectively design solutions for communal problems. In such a model, experts play a role but do not retain exclusive power over decision making; typically, experts play a facilitator role in the deliberative process (Fung and Wright 2001).

Participation Design Matters

Carrie Menkel-Meadow (2002) elaborates on the importance of facilitators in the deliberative process. As she notes, decision making in participation forums is distinguishable from more conventional forms of decision making, such as command-and-control, decision making by experts, aggregative voting, and strategic negotiation. In expert decision making, discussions are segregated from popular participation and often neglect viewpoints. In aggregative voting, participants simply vote according to their own self-interest, without considering the reasonableness, fairness, or acceptability of each option. In the strategic negotiation model, Menkel-Meadow (2002) notes that decision making procedures use bargaining strategies to promote participants' own self-interest backed by resources and power. Deliberative decision making, however, is animated by commitment to practices and values of communication, public justification, and deliberation (Menkel-Meadow 2002).[3] To ensure genuine participation by all as opposed to only those who have the advantages of wealth, education, or membership in dominant groups, Menkel-Meadow suggests that a "neutral third party" play a procedural role in the deliberative process. In particular, such a third party can map potential interests and stakeholders, assist with the design and implementation of the deliberative process and decision rules, and resolve conflicts if necessary (Menkel-Meadow 2002, 76).

Furthermore, Menkel-Meadow (2011) contends that strong procedural rules are crucial to ensure that all sides have a genuine opportunity to influence the process and its outcomes, particularly in highly controversial arenas involving multiple stakeholders and conflicting sides. Procedural rules include rules for agenda setting, argumentation, justification, counterargumentation, and deliberation. Menkel-Meadow recalls that most of the current understanding concerning political participation is rooted in Habermas's ideal speech conditions for deliberation, which are reasoned argument, uncoerced speech, equal opportunities, rational persuasion, and openness to hearing other participants' opinions (Habermas 1996). In real-life deliberation, however, Habermas's ideal conditions consider only one mode of discourse (principled-rational) and overlook two other types of discourse: the bargaining or preference-trading discourse (negotiation), and the affecting and value-feeling discourse (emotions-, religious-, and consciousness-based discourse) (Menkel-Meadow 2011). Thus, deliberative forums need to be carefully designed to deal with all three forms of discourse and ensure fair procedures for resolving disagreements.

Lessons for Policy Making in Agricultural Genomics in Canada: The European Union's Water Framework Directive

There are many ongoing experiments attempting to foster participation of ordinary people in policy decision making.[4] This section discusses the European Union's Water Framework Directive (WFD) as a promising example for Canada in terms of improved democratic policy making and better coordination of multilayered legal frameworks.

In October 2000, the Council of the European Parliament adopted a legal *Framework for Community Action in the Field of Water Policy* (Water Framework Directive, or WFD).

The WFD resulted from a general dissatisfaction with the traditional governance methods of EU environmental law in general, and water quality in particular. In the late 1990s, a series of eleven separate water directives were in place, yet stakeholders contended that the frameworks failed to take into account the distinctive conditions and interests of the various member states, along with European Union's limited capacity to ensure compliance (Holzinger, Knill, and Schäfer 2006). Similar concerns appear to be present in the legal agricultural genomics landscape in Canada, particularly multilayered legal frameworks with divergent approaches (see Chapter 1) and disregard for distinctive interests (see Chapter 3). With the objective of replacing the series of multilayered, ineffective regulations and top-down rules governing water issues in the European Union, the WFD proposes a combination of general binding requirements with more specific nonbinding standards and objectives to address distinctive needs. In this way, the WFD combines the old and new types of governance: detailed rules binding on member states, subject to judicial enforcement and formal sanctions, but also open-ended guidelines and direct stakeholder participation in both nonbinding guidelines and binding rules (Trubek and Trubek 2007, 13).

As Trubek and Trubek (2007, 16) summarize, member states are required to avoid any deterioration of water quality and attain "good water status" in a given timeframe, and a supplementary addendum provides thorough guidelines on what constitutes "good water status" for specific types of water. There are no final and uniform standards for measurement; rather, standards are subject to further development by stakeholders through a series of collaborative participation processes to ensure that member states and the European Commission are involved in a collaborative and coordinated fashion.

In fact, rather than a piece of legislation, the WFD is a programmatic set of rules and guidelines that aim to ensure democratic policy making along the process. To this end, Article 14 of the WFD expressly requires member states to "encourage the active involvement of all interested parties in the implementation of this Directive." To be sure, Article 14 encourages participation in the implementation phase particularly because most of the work is realized then; in reality, however, public participation is required at all stages.[5] To ensure that the regulatory tasks and executive responsibilities are implemented on an ongoing basis, the European Commission and member states created a central organizational committee, the Common Implementation Strategy (CIS). The CIS employs a horizontal network system made up of representatives of member states, experts, and members of the public who work collaboratively to design indicators to measure progress and processes for information sharing and management plans, and to produce nonbinding guidance documents.

The WFD seems to illustrate a promising mixed model of old and new forms of governance that permits a diversity of perspectives and problem-solving initiatives but also holds stakeholders to measurable results and formal sanctions. By involving ordinary people and state actors in a collaborative effort to solve specific problems, by relying on deliberative decision making in which participants hold real decision-making power, and by creating formal linkages of coordination and information sharing as part of the state institutional apparatus, the Water Framework Directive in general, and the Common Implementation Strategy in particular, appear to be a promising experiment toward more democratic policy-making institutions. However, echoing Trubek and Trubek's concerns (2007), more research is needed to identify whether and the extent to which the ambitious goals of increased democratic policy making, responsiveness to different interests, and enhanced legal coordination have in fact been achieved. Nonetheless, the WFD model can serve as a starting point for Canadian agricultural genomics policy makers and researchers to reflect on ways to better coordinate multilayered legal frameworks and divergent viewpoints in an effective and democratic fashion. The domestic and international IP–Regulatory Complex governing agricultural genomics research is massive and tricky, and the net impact on the development and delivery of genomics research to end-users is yet to be determined (see Chapter 1). The lack of coordination not only between agencies but also between groups concerned about the issue seems to be a shared grievance among stakeholders and researchers. By using a participatory governance model, agencies and all stakeholders have a voice to communicate and justify their positions. Not only can this prevent conflicts and duplication of regulations but it also allows for rapid adaptation of policy during periods of technological change. Although, as part of the current process, the Senate invites parties to speak about their concerns, the Senate makes a one-time recommendation, which may or may not be acted on by the legislature and executive.[6] In addition, the process is only a consultation or clarification process in which people who speak before the Senate have no decision-making power, and their voices may or may not be taken into account in the Senate's final recommendations. A structure such as the Water Framework Directive in the European Union and the institutional and procedural arrangements proposed by Fung and Wright and Menkel-Meadow would become fluid, allowing for democratic decision making and agile adaptation of new technologies and issues into policy.

A major change is taking place in the way governments make decisions and achieve public goals. The traditional state-centred landscape within which law- and regulation-making processes were devised is being reconfigured into a contemporary model of governance that engages state and nonstate actors in policy making. In the governance model, these actors cooperate in designing flexible norms, standards, benchmarks, and processes of peer review to ensure time- and context-sensitive norms and accountability. The governance model emerged largely due to the failure of regulatory models to adequately address the complexities of society's current social dilemmas. Proposals to facilitate greater efficiency in meeting the aims of intellectual property and regulatory regimes while enabling innovative research must take into account these fundamental changes in political and societal affairs. In light of these factors, this chapter suggests that a governance approach to the IP–regulatory conundrum and wider stakeholder participation in the development of solutions based on flexible and revisable rules might be better suited to ensure time- and context-sensitive norms and accountability. There are many governance experiments based on a wide range of configurations that can serve as guidelines to help Canadians structure a multilevel govern-ance system. The European Union's Water Framework Directive is but one of the existing models.

Again, it is important to keep in mind that this chapter does not suggest that a governance approach will dissolve the barriers that impede the Intel-lectual Property–Regulatory Complex in meeting the goal of delivering innovative research to end-users. This chapter only acknowledges the on-going changes in political institutions and decision-making processes, and attempts to connect these fundamental changes to the agricultural genom-ics landscape. The hope is that applying a different paradigm to the issues and barriers posed by the IP–Regulatory Complex might shed some light on how to bring greater pluralism and efficiency to bear, and facilitate the ultimate goal of delivering innovative genomics research to end-users in Canada.

Finally, it is hoped that this chapter will incite a more in-depth inquiry into the significance and effectiveness of a governance approach to the agricultural genomics landscape. Further inquiry might include a critical analysis of the failures of the regulatory model to deal with the challenges of the agricultural genomics complex; an examination of the feasibility of a governance approach in light of Canada's legal system; and more adequate participation designed to foster state and nonstate actors in deliberative de-cision making in the agricultural genomics research arena.

Notes

1 A new approach to governance is not necessarily a "new" form of policy making but it is a new form of policy making in relation to the traditional forms of policy making based on state actors and centralized policy making.
2 Several studies examine the relationship between regulatory and governance ap-proaches, discussing the successes and failures of a wide range of configurations. See, e.g., Scott and Trubek 2002; Trubek 2006; Trubek and Trubek 2007, 2010.
3 See also Fung and Wright 2001.
4 This discussion of the example of the Water Framework Directive builds primarily on David Trubek, Louise Trubek, and Joanne Scott's work; see, e.g., Scott and Trubek 2002. "Ongoing experiments": see, e.g., Fung and Warren 2011, discussing the *Participedia* Project, an online-based collection of participation experiments around the globe.
5 See European Commission 2003, a detailed guidance document to assist with the implementation of public participation requirements.
6 See, e.g., Standing Senate Committee on Agriculture and Forestry 2012, setting out the evidence on research and innovation efforts in the agricultural sector, particu-larly the role of intellectual property rights for innovation in agriculture.

References

Legislation

EC, *Council Directive 2000/60/EC of 23 October 2000 Establishing a Framework for Community Action in the Field of Water Policy,* 2000 O.J. L 327/1 [Water Framework Directive, or WFD].

Secondary Sources

Arnstein, Sherry R. 1969. "A Ladder of Citizen Participation." *Journal of the American Institute of Planners* 35 (4): 216–24. http://dx.doi.org/10.1080/01944366 908977225.

Ayres, Ian, and John Braithwaite. 1992. *Responsive Regulation: Transcending the Deregulation Debate.* New York: Oxford University Press.

Baldwin, Robert, Martin Cave, and Martin Lodge. 2012. *Understanding Regulation: Theory, Strategy and Practice.* 2nd ed. Oxford: Oxford University Press.

Barber, Benjamin R. 1984. *Strong Democracy: Participatory Politics for a New Age.* Berkeley: University of California Press.

Costa, N. 1986. *Lutas Urbanas e Controle Sanitário: Origens das Políticas de Saúde no Brasil.* Petrópolis: Vozes.

Edelstein, Michael R. 1988. *Contaminated Communities: The Social and Psychological Impacts of Residential Toxic Exposure.* Boulder, CO.: Westview Press.

European Commission. 2003. *Public Participation in Relation to the Water Framework Directive.* Guidance Document No. 8. Luxembourg: European Commission – Communication and Information Resource Centre for Administrations, Businesses and Citizens Library. https://circabc.europa.eu/sd/a/0fc804ff-5fe6 -4874-8e0d-de3e47637a63/Guidance%20No%208%20-%20Public%20 participation%20%28WG%202.9%29.pdf%20b.

Fischer, Frank. 1993. "Citizen Participation and the Democratization of Policy Expertise: From Theoretical Inquiry to Practical Cases." *Policy Sciences* 26 (3): 165–87. http://dx.doi.org/10.1007/BF00999715.

Fischer, Robert. 2008. "European Governance Still Technocratic? New Modes of Governance for Food Safety Regulation in the European Union." In *European Integration Online Papers,* vol. 12. http://ssrn.com/abstract=1339831.

Fung, Archon, and Mark E. Warren. 2011. "The *Participedia* Project: An Introduction." *International Public Management Journal* 14 (3): 341–62. http://dx.doi. org/10.1080/10967494.2011.618309.

Fung, Archon, and Erik Olin Wright. 2001. "Deepening Democracy: Innovations in Empowered Participatory Governance." *Politics and Society* 29 (1): 5–41. http:// dx.doi.org/10.1177/0032329201029001002.

Funtowicz, Silvio O., and Jerome R. Ravetz. 1993. "Science for the Post-Normal Age." *Futures* 25 (7): 739–55. http://dx.doi.org/10.1016/0016-3287(93)90022-L.

Habermas, Jürgen. 1970. "Technology and Science as 'Ideology.'" In *Toward a Rational Society: Student Protest, Science and Politics,* trans. J. Shapiro. Boston: Beacon Press.

–. 1996. *Between Facts and Norms: Contributions to a Discourse Theory of Law and Democracy.* Cambridge: Polity Press.

Hage, Maria, Pieter Leroy, and Elmar Willems. 2006. "Participatory Approaches in Governance and in Knowledge Production: What Makes the Difference?" Working Paper Series 2006/3. Research Group on Governance and Places, University of Nijmegen. http://repository.ubn.ru.nl/bitstream/handle/2066/46439/46439.pdf?sequence=1.

Hart, Andy, ed. 2003. *Improving the Interface between Risk Assessment and Risk Management.* Final Report of a European Workshop on the Interface between Risk Assessment and Risk Management, Noordwijkerhout, Netherlands, September 3–5, 2003. Sand Hutton, York: Central Science Laboratory.

Holzinger, Katharina, Christoph Knill, and Ansgar Schäfer. 2006. "Rhetoric or Reality? 'New Governance' in EU Environmental Policy." *European Law Journal* 12 (3): 403–20. http://dx.doi.org/10.1111/j.1468-0386.2006.00323.x.

Johnson, James. 1998. "Arguing for Deliberation: Some Skeptical Considerations." In *Deliberative Democracy,* ed. Jon Elster. New York: Cambridge University Press. http://dx.doi.org/10.1017/CBO9781139175005.009.

Kooiman, Jan. 2003. *Governing as Governance.* London: Sage.

Lobel, Orly. 2004. "The Renew Deal: The Fall of Regulation and the Rise of Governance in Contemporary Legal Thought." *Minnesota Law Review* 89: 262–390.

Menkel-Meadow, Carrie. 2002. "The Lawyer as Consensus Builder: Ethics for a New Practice." *Tennessee Law Review* 70 (1): 63–119.

–. 2011. "Scaling Up Deliberative Democracy as Dispute Resolution in Healthcare Reform: A Work in Progress." *Law and Contemporary Problems* 74 (3): 1–30.

Richardson, Mary, Joan Sherman, and Michael Anthony Gismondi. 1993. *Winning Back the Words: Confronting Experts in an Environmental Public Hearing.* Toronto: Garamond Press.

Salamon, Lester. 2001. "The New Governance and the Tools of Public Action: An Introduction." *Fordham Urban Law Journal* 28 (5): 1611–74.

Sanders, Lynn M. 1997. "Against Deliberation." *Political Theory* 25 (3): 347–76. http://dx.doi.org/10.1177/0090591797025003002.

Scott, Joanne, and David Trubek. 2002. "Mind the Gap: Law and New Approaches to Governance in the European Union." *European Law Journal* 8 (1): 1–18. http://dx.doi.org/10.1111/1468-0386.00139.

Standing Senate Committee on Agriculture and Forestry. 2012. *Evidence.* 41st Parliament, 1st Session. October 25, 2012. http://www.parl.gc.ca/content/sen/committee/411/AGFO/49755-e.HTM.

Trubek, David M., and Louise G. Trubek. 2007. *New Governance and Legal Regulation: Complementarity, Rivalry or Transformation.* University of Wisconsin Legal Studies Research Paper No. 1047. Social Science Research Network, http://papers.ssrn.com/sol3/papers.cfm?abstract_id=988065.

–. 2010. "Afterword – Part I: The World Turned Upside Down: Reflections on New Governance and the Transformation of Law." *Wisconsin Law Review* 2: 719–26.

Trubek, Louise. 2006. *New Governance and Soft Law in Health Care Reform.* University of Wisconsin Legal Studies Research Paper No. 1018. Social Science Research Network, http://papers.ssrn.com/sol3/papers.cfm?abstract_id=908674.

Tushnet, Mark. 2003. *The New Constitutional Order.* Princeton, NJ: Princeton University Press.

8

Constructing an International Intellectual Property *Acquis* for the Agricultural Sciences

ROCHELLE COOPER DREYFUSS

The agricultural industries have increasingly come to rely on intellectual property law to protect investments in improved plants and breeding techniques. While these appropriation strategies (along with new self-help opportunities) enhance profits and increase incentives to invest, they can also endanger innovation, contribute to an unjust distribution of knowledge outputs, and lead to practices that undermine food security. At one time, every country could balance the costs and benefits of intellectual property on its own and craft law that met its individual interests. As global trade has increased, however, national flexibilities have diminished. Interstate competition means that every country must consider what its neighbours are doing. As important, intellectual property lawmaking has moved to the international level. Negotiations have been characterized by regime shifts and successively more strident demands for ever-stronger protection. Weaker countries, along with other proponents of a more open technological environment, have had a difficult time being heard. As Daniel Gervais has informally said, "intellectual property is on a grain elevator: it moves only in an upward direction." Demonstrations in Doha, in Seattle, in Wellington – wherever international negotiators meet to discuss intellectual property issues – suggest that a new approach, one that is more attentive to broader social values, is necessary (Kapczynski 2008; TPP Watch 2012).

After discussing the nature of the problems confronting the agricultural sector and the difficulties states encounter in solving these problems domestically, this chapter applies a framework I developed with Graeme Dinwoodie in our recent book, *A Neofederalist Vision of TRIPS: The Resilience of the International Intellectual Property Regime* (Dinwoodie and Dreyfuss 2012). In the book, we argue that when intellectual property law is viewed holistically and over time, there are certain core principles that appear repeatedly. While many of these principles protect rights holders and state interests, there are also principles that protect public access and future generations of innovators. These are harder to find in international conventions. Negotiators know more about trade law than intellectual property law, and they are heavily influenced by the creative industries. The result is that these agreements may leave some room for public-regarding measures, but the details are left to member states. It is thus the job of scholars to identify these principles and explain their critical role in innovation policy to legislators, courts, negotiators, and international adjudicators. In the book, Professor Dinwoodie and I considered the appropriate balance in the context of the arts and sciences. This chapter takes a closer look at agricultural innovation.

The Problem

As recounted by Glenn Bugos and Dan Kevles (1992), the last century has witnessed relentless pressure to increase intellectual property protection for advances in the agricultural sector. US history is illustrative. At the outset, innovation in this arena was largely financed by the government and made freely available to farmers. When private plant breeding became more prevalent toward the end of the nineteenth century, breeders initially had few tools to combat piracy. They could lock seeds and samples in safes, charge high prices for the first generation of a new plant, and contractually obligate buyers to refrain from reseeding or grafting. However, monitoring workers, policing farms, and enforcing contracts were costly endeavours and not always fully effective. Unauthorized breeding – and fraud – were constant problems, making federal intellectual property law highly desirable. In 1881, federal trademark law was enacted (*Trade-mark Act*). It gave breeders the ability to use signifiers to help consumers find and appreciate the quality of their goods, but it did not directly protect either plants or breeding techniques. The legal situation changed dramatically in the mid-twentieth century. In 1930, the *Plant Patent Act* created breeders' rights in asexually reproduced plants, and in 1970, the *Plant Variety Protection Act* accorded protection to plants that reproduce sexually. With the US Supreme Court's 1980 decision in *Diamond v. Chakrabarty* holding manmade microorganisms to be eligible for utility patents, plants became subsumed into the general patent system.[1] Utility patents are now used to protect not only plants but also individual genes, proteins, and pathways of development and metabolism, along with methods of production, selection, and analysis (Louwaars et al. 2009).

Breeders have come to enjoy a variety of other protections as well. As the enforcement of non-negotiated agreements (variously known as shrinkwraps, clickwraps, and bag-tag licences) became judicially accepted, breeders could use these instruments to easily bind buyers to restrictions on reseeding, transferring seeds, or conducting research (Rowe 2011; *ProCD v. Zeidenberg*[2]). Some countries (not the United States) provide protection for data compilations, which can include information relevant to farming.[3] Technological advances have created new opportunities as well. Among other things, breeders can use "terminator" technology to render plants infertile and unavailable as stock for future growing programs (Muscati 2005).[4]

In many ways, these new strategies are salutary developments. They enhance appropriability and thus increase incentives to invest in agricultural innovation. In a world with a growing population and potentially dramatic climate change, there is a clear need for a sustained focus and more research on such matters as plant productivity, resistance to disease, and tolerance to environmental conditions. At the same time, however, strong intellectual property rights can also endanger innovation. Knowledge is cumulative, especially in the agricultural sector, and is heavily dependent on access to existing stock and genetic materials, to a specific set of techniques, and to the empiricism of farmers and researchers.

From a distributive justice perspective, these developments are also worrisome. Intellectual property protection raises costs without spurring innovations uniquely needed by nonmarket economies or small farmers (Mechlem 2010). Agricultural scientists might do such work for humanitarian or reputational reasons, but, without profits, paying for the inputs gets more difficult as prices rise. Furthermore, the changing legal environment has been accompanied by significant mergers and aggressive acquisitions in the breeding industry (Strauss 2009). With fewer research programs, technologically superior seed choices become more limited, and monocultures become more likely. These endanger food security, and as Ron Herring has shown, farmers may find that no choice is congenial to their own local environments. For further discussion, see Ron Herring's discussion of stealth seeds in Chapter 4.

In the initial foray into intellectual property protection, there was minimal danger that one generation of rights would block future generations of innovators or have an impact on nutrition. For example, the *Plant Patent Act* excludes tuber-propagated plants because the reproductive element is also a food (Bugos and Kevles 1992).[5] The *Plant Variety Protection Act* creates rights over single varieties of a plant and is subject to exemptions favouring farmers and researchers.[6] It also contemplates the award of compulsory licences in order to ensure an adequate food supply.[7] The newer protections are far less modest. Utility patents cover all types of plants (so long as they are inventive and adequately disclosed) and do not include crop exemptions; there are no compulsory licensing provisions relevant to agriculture (*Dawson Chemical Co. v. Rohm and Haas Co.*).[8] The US Court of Appeals for the Federal Circuit, which hears virtually all patent appeals, has severely limited what had been a long-standing research defence (*Madey v. Duke Univ.*), and the US Supreme Court has curtailed antitrust scrutiny of refusals to deal (*Verizon Comm'ns Inc. v. Law Offices of Curtis V. Trinko, LLP*).[9]

To make matters worse, some of these patents can be extremely broad. Gene patents furnish examples. Thus, in *Association for Molecular Pathology v. USPTO* – the "*Myriad*" case on the patentability of BRCA mutations associated with early-onset breast cancer – US Federal Circuit Judge Bryson described one of the claims as follows (at 1376, Bryson, J., dissenting in part):

> The patent contains a sequence that is ... 24,000 nucleotides long with numerous gaps denoted "vvvvvvvvvvvvvv." An almost incalculably large number of new molecules could be created by filling in those gaps with almost any nucleotide sequence, and all of those molecules would fall within the scope of [the] claim.

While the Supreme Court failed to opine on that aspect of the case in its opinion in *Ass'n for Molecular Pathology v. Myriad Genetics, Inc.*, the Court appears to have approved another broad claim: one that covers any cDNA sequence that has as few as fifteen nucleotides in common with a described full-length sequence (at 2113, describing claim 6 of US Patent 5,747,282). The breadth of these patents makes it difficult to determine freedom to operate. Nor is it clear that those holding them have made contributions commensurate with the reward these rights make available (Nuffield Council 2002). Furthermore, some genetic material is not variety-specific. For example, a patent on genetic information in one cereal may, in fact, describe and cover all cereals. Genes have even been known to jump species.[10] As a result, before research can be done, permissions may be needed from rights holders in fields far removed from the one in which a particular breeder is operating. As more genes, techniques, and other outputs are protected, transaction costs rise, the opportunity for holdouts increases, and royalties begin to stack to the point where some research – especially risky research, research right at the cutting edge – is less likely to be undertaken (Huang and Murray 2009). As Rebecca Eisenberg has demonstrated in her work on bargaining in the biotech sector, heterogeneity among different rights holders means that their expectations and demands can be quite different, leading to complicated and lengthy – or failed – negotiations (Eisenberg 2001, 223).

The Limits of Domestic Solutions

Ostensibly, each nation could deal with the adverse impact of intellectual property rights by simply changing its laws to rebalance interests in light of new developments and local conditions. For example, in the United States, the Supreme Court has indicated in *Mayo Collaborative Serv. v. Prometheus Laboratories, Inc.* that natural laws cannot be patented unless the inventor adds more to the claims than "well-understood, routine, conventional activity previously engaged in by scientists in the field," and in *Myriad* the Supreme Court drew a line between gDNA, which it regarded as an unpatentable phenomenon of nature and cDNA, which it saw as manmade and therefore patentable. Similarly, the utility requirement has been interpreted in a manner that ensures that a patent applicant has contributed significantly to the field (*In re Fisher*). And as Judge Bryson intimated in his lower court opinion in the *Myriad* case, the enablement and written description requirements may limit the scope of some gene patents.[11] Farmers are also seeking (so far unsuccessfully) to have the courts define infringement in a way that would protect them from liability if patented seeds inadvertently grow on their land (*Organic Seed Growers and Trade Ass'n v. Monsanto Co.*).[12] Scholars have recommended other approaches, including the adoption of defences akin to the fair use defence of copyright law and antitrust scrutiny of postsale restraints (Strandburg 2011; O'Rourke 2000; Carstensen 2006).

Other countries have adopted analogous strategies. In Germany and France, the scope of gene patents is limited to the utility disclosed in the patent application.[13] In *Monsanto Tech. LLC v. Cefetra BV,* the Court of Justice of the European Union (CJEU) held that a patent on a gene in a soy plant was not infringed by meal made from the grown plant. In *Cefetra,* the court reasoned that in the feed, the gene was no longer functioning as described in the patent. And in *Radio Telefis Eireann & Indep. Television Publ'ns Ltd. v. Comm'n of the European Communities,* the CJEU denied a rights holder who lacked business justification the power to refuse to license the copyright to a party who would bring to market a product consumers desired (paras. 53 to 54).[14]

Unfortunately, however, states suffer from a collective action problem. Because trade is global, the location of a firm's headquarters, manufacturing facilities, or research and development laboratories can be chosen in light of the applicable law (O'Brien 2012). Since an industrial presence creates jobs and wealth, nations are under considerable pressure to adopt law that meets industry's demands. The Canadian experience is instructive. In *Harvard College v. Canada (Commissioner of Patents),* the Supreme Court of Canada denied utility patent protection to Harvard's oncomouse. The reach of the decision was such that it might have stopped utility patenting of living things in its tracks. But there was a long dissent by Justice Binnie, who worried about Canada's competitive position vis-à-vis the United States (para. 13): "The mobility of capital and technology makes it desirable that comparable jurisdictions with comparable intellectual property legislation arrive at similar legal results."

Indeed, the majority rather quickly came around to the dissenting justice's view; in *Monsanto Canada Inc. v. Schmeiser,* the Court upheld gene and cell patents associated with Roundup Ready canola (and did not, apparently, require proof that the alleged infringer had made use of the patented technology by applying the herbicide Roundup to his crops).

Importantly, states are under escalating international pressure to provide strong protection to agriculture. Initially, international requirements grew apace with domestic law, but in recent years the impetus to legislate at the international level has accelerated. These efforts are characterized by intensive regime shifting – moves between subject matter–based organizations and generalized forums, and among multilateral organizations (the World Intellectual Property Organization [WIPO] or the World Trade Organization [WTO]), bilateral discussions, and plurilateral negotiations. Failed efforts in one forum reappear in another, where the national lineups are different, in the hope that leverage will change so that proposals that failed in one setting will succeed in another. Thus, while the nineteenth-century intellectual property agreements the *Paris Convention for the Protection of Industrial Property* (Paris Convention) and the *Berne Convention for the*

Protection of Literary and Artistic Works (Berne Convention), had little to say about plants, 1961 saw the adoption of the *International Convention for the Protection of New Varieties of Plants* (UPOV Convention), which requires protection for plant varieties that are new, distinct, uniform, and stable. The UPOV Convention was not, however, able to attract more than seventy members to recognize these plant breeders' rights.

Many more countries belong to WIPO, but WIPO has never managed to impose a requirement to protect plants on its membership. During the Uruguay Round, however, developed countries reconceptualized intellectual property as a trade issue. The *Agreement on Trade-Related Aspects of Intellectual Property Rights* (TRIPS Agreement), which came into force in 1995, requires each of the 155 members of the WTO to "provide for the protection of plant varieties either via patents or by an effective *sui generis* system or by any combination thereof" (Article 27.3[b]). It also requires protection for geographical indications of foodstuffs as well as protection against unfair competition for undisclosed information, and, in particular, for data on agricultural chemical products that are submitted to regulatory agencies (Articles 22–24, 39.1, and 39.3).

Despite the success of the TRIPS Agreement, it too was not the end of the road. Developing nations with diverse ecological resources and strong traditions on how plants can be used are now seeking compensation for the use of these endowments. Thus, the *Nagoya Protocol* to the *Convention on Biological Diversity*, which was adopted in October 2010, will require those who wish to use genetic resources commercially or for research and development to obtain consent from the state where the resources are found. WIPO is similarly working on protocols to protect traditional knowledge, genetic resources, and traditional cultural expression.[15]

Meanwhile, developed countries have been active as well. The United States (aided, to some extent, by other countries) is engaged in negotiating bilateral agreements containing TRIPS-plus commitments. For example, the United States has entered into free trade agreements requiring UPOV Convention protection and/or utility patent protection for plants; examples include the *United States–Morocco Free Trade Agreement*, Article 15.9(2) (a), and the *United States–Chile Free Trade Agreement*, Articles 17.2 (3)(a) and 17.9(2). There is also a move to include such obligations in plurilateral arrangements. Thus, the *Anti-Counterfeiting Trade Agreement* (ACTA) would require stricter border and remedial measures, which may affect the usage of patented plant resources (Yu 2011). According to leaked texts of the proposed *Trans-Pacific Partnership* (TPP),[16] parties to that agreement will be required to ratify the UPOV Convention (Article 1.3[g]), recognize utility patent protection for plants (Article 8.2[a]), and protect premarket clearance data regarding agricultural chemical products for a period of ten years from the date of marketing approval in the territory (Article 9.1). Although ACTA and the TPP involve a limited number of parties, it is already clear that if they come into force, other countries will be pressured to join. For instance, Mexico and Canada were initially not involved in either initiative; now both have joined ACTA and have become involved in TPP negotiations (*Anti-Counterfeiting Trade Agreement;* Foreign Affairs, Trade and Development Canada [see p. 231] 2013).

Admittedly, there are countermoves afoot. Organizations interested in human rights, development, food security, and health have explored the flexibilities remaining under the agreements and have helped governments find ways to work with them (Helfer 2004).[17] At the end of the day, however, those who wish to explore existing plant resources could come to need permissions from the holders of breeders' rights, patent rights, and data rights, as well as authorization from specific nations and perhaps even indigenous groups (or individual farmers or tribal leaders) who are located at the source of genetic or phenotypic information. Any one of these parties could refuse to authorize use and hold up research or the exploitation of important advances. Furthermore, because there is no hierarchy among international lawmaking authorities, disputes over intellectual property rights could easily cycle, with tribunals reaching different – and conflicting – decisions over the allocation of rights in new knowledge products.[18] Significantly, the proposed agreements include few defences or exceptions to protect access interests. Nor do they provide mechanisms to avoid cycling.

A New Approach

Is there anything that can be done to preserve national flexibilities to deal with escalating demands for the protection of agriculture? Professor Dinwoodie and I have suggested that the international system must develop ways to avoid legal fragmentation and to consider the benefits and costs of intellectual property protection in a more systematic fashion. One set of proposals deals with governance issues. These include urging international organizations to allocate decision-making responsibility as part of their lawmaking efforts, as well as the recognition of national norms to restrain regime shifting, concepts akin to *forum non conveniens* and *res judicata* to prevent one regime from undermining the choices made in another, choice of law rules to avoid conflicting obligations, and global administrative norms that promote open deliberation on international instruments and provide opportunities for all of those affected to be heard.

These ideas would be helpful in any area of international law. We think, however, that more is necessary in the intellectual property arena. In that domain, the salience of knowledge products in the modern economy has made demands uniquely incessant. Moreover, the match-ups between the demandeurs of strong protection and those resisting their efforts are persistently asymmetrical, with rights holders much more organized and better positioned to influence negotiations. In addition, because these negotiations mostly take place in a trade context, the negotiators are often not versed in intellectual property law or lore, and they are so focused on the balance of *payments* that they may not fully understand the need for balance among *users and producers*. Increasingly, these agreements are also negotiated in secret, meaning that the negotiators are largely shielded from information they need to hear. The TPP negotiations furnish a good example. They have been taking place entirely in secret. Significantly, when US academics asked for greater transparency and expanded participation, the US Trade Representative claimed that comments had already been solicited from "a wide range of stakeholders" (Infojustice 2012). Since public interest groups have not, on the whole, been asked their views, it is clear that those interested in the public domain or robust access safeguards are not considered stakeholders.[19] With that definition in place, it is unlikely that the public's position will be fully considered in framing new agreements.

There is an additional problem: international intellectual property law has long been devoted to creating rights. Exceptions are framed as merely permissible. That structure has several ramifications. Legislatures obliged to execute international commitments focus on what they must do; permissible exemptions can fall by the wayside. In many legal systems, rights are construed expansively but exceptions are interpreted narrowly – users can wind up with less than they need; producers get more than they deserve. Because the rationale for exceptions is poorly articulated, international adjudicators have little authoritative material to rely on when deciding whether a particular national exception is allowable. Given the unpredictability of outcomes, states may be less willing to promote access interests. The agreements, in effect, enhance the leverage of those asking for stronger protection and diminish the clout of anyone who is opposed.

To cope with these difficulties, Professor Dinwoodie and I propose the development of what we have called an international intellectual property *acquis*. We draw the idea from the European Union and the WTO. Both bodies recognize that in addition to the specific measures set out in the constitutive instruments, there is an underlying body of principles that are part of the regime, even though unexpressed (Gadbaw 2010, 567).[20] These principles further legitimate expectations, fill gaps, and create certainty and predictability. Articulating an *acquis* for intellectual property would likewise make explicit the structure of the intellectual property system as a whole. In our book, we examine not only the express provisions of international instruments but also principles that appear repeatedly in national laws, doctrines that undergird long-standing national practices, values embedded in cognate bodies of law, and norms reflected in national constitutions. These "multisourced equivalent norms" (to use a term coined by Tomer Broude and Yuval Shany) demonstrate a consensus among policy makers, which, we argue, ought to be taken into account in every lawmaking effort (Broude and Shany 2011). The *acquis* as a whole would include principles protecting those who develop intellectual property, and state interests as well. But the major significance of the *acquis* would be to bring to the fore those principles that protect the consumers of intellectual material and use it to push the frontiers of knowledge outward.

Explicit recognition of user interests would have many advantages. It would provide states with guidelines on what is possible in light of their international obligations and add normative heft to pro-access arguments. It would also put access safeguards on a level with the protection afforded to rights holders. In emerging areas (such as biotech-based agricultural innovation), the *acquis* would provide guidance on where and how to strike the optimal balance between public and private interests. The *acquis* could also offer adjudicators a better handle on all sides of the intellectual property equation and facilitate dispute resolution. Most important, it would create a legal framework for structuring future international lawmaking and, hopefully, take intellectual property law off the grain elevator so that it is no longer moving ever upward.

Some may be skeptical about the notion of principles that transcend individual instruments, nations, or organizations. But as the Westphalian state has ceded control to a complex array of international authorities, the need for a new conception of governance has become increasingly evident. Benedict Kingsbury, Nico Krisch, and Richard Stewart (2005) have advocated for the development of global administrative law, but administrative norms are not sufficient when negotiators define interested stakeholders narrowly. In some areas, *jus cogens* is a strong limiting force.[21] However, there has never been a *jus cogens* of intellectual property law – in a sense, the *acquis* is prompted by the need to create an analogous concept. Eyal Benvenisti and George Downs (2012) suggest that national courts might provide checks and balances to international lawmaking, but without a recognized set of core principles, the collective action aspect of the problem makes that approach ineffective. Finally, as Alison Slade has argued, on the user side, the project may have already begun. The Objectives and Principles of the TRIPS Agreement were prominently mentioned in the *Doha Ministerial Declaration* and the *Declaration on TRIPS and Public Health*, they have been relied on in WIPO considerations of its Development Agenda, and are referenced in ACTA (Slade 2011).[22] Their provisions can therefore be used to ground our project.

What, then, are the principles safeguarding access interests? Looking at intellectual property laws holistically, there are three types of strategies: subject matter exclusions, scope limitations, and laws that control abuse.

Subject Matter Exclusions
Even the major conventions recognize that some absolute exclusions are necessary. For example, the Berne Convention excludes from copyright protection "news of the day or ... miscellaneous facts having the character of mere items of press information" (Article 2[8]). Likewise, the TRIPS Agreement excludes "ideas, procedures, methods of operation or mathematical concepts as such" (Article 9.2).[23] And in the patent realm, the TRIPS Agreement permits states to exclude "diagnostic, surgical, and therapeutic methods, plants (with a proviso concerning the availability of *sui generis* rights), and any invention necessary to protect public order, life, health, or the environment (Articles 27.2 and 27.3). National courts do not often interpret international instruments, but those that have suggest that they function as ceilings on protection. For example, in *CCH Canadian v. Law Society of Upper Canada*, the Supreme Court of Canada suggested that the Berne Convention imposes an originality requirement in copyright.[24]

National legislation is even clearer about the notion that some advances, even if they are new to the storehouse of knowledge, cannot be protected: ideas and facts, generic terms, principles and phenomena of nature, abstract ideas, and mathematical methods and formulae. As the US Supreme Court put it in *Bilski v. Kappos* at 3225: "The concepts covered by these exceptions are 'part of the storehouse of knowledge of all men ... free to all men and reserved exclusively to none.'" Some national laws are more explicit. They exclude plants and animals, diagnostic methods, and inventions necessary to safeguard the public order (Bently et al. 2010; *European Patent Convention,* Articles 52 and 53). Or they have raised the inventive step to a level high enough to permit a degree of tinkering and to prohibit evergreening.[25]

Together, these provisions suggest that the basic building blocks of creativity, the fundamental components of expression, and elements essential to vigorous competition must remain in the public domain. In other words, they demonstrate a consensus that nations can act to maintain a competitive innovation environment, reduce transaction costs, and ensure that authors and inventors have contributed enough to society to merit the protection the law accords them (Kaplan 1967).

Scope Limitations

Intellectual property regimes are equally careful to police the limits of the protection they provide. For example, the Berne Convention includes a mandatory quotation right and a set of specific exemptions for teaching and reporting (Articles 10[1], 10[2], and 10bis). The TRIPS Agreement permits states to revoke trademarks when they lose distinctiveness (Articles 19 and 21), and to issue compulsory patent licences to meet local needs, to protect health, safety, and nutrition, and to prevent one patent from blocking the exploitation of another (Article 30). Further, the TRIPS Agreement allows judges to take the public interest into account when framing remedies (Articles 31[b] and 44.2). Significantly, all these instruments include general exceptions to give signatories freedom to constrain the reach of intellectual property rights in other ways.[26]

National laws have adopted several strategies for using these flexibilities. They limit the term of intellectual property rights – indeed, in the United States, the limit is a constitutional requirement.[27] On the whole, countries recognize a doctrine of local (if not international) exhaustion, which gives purchasers the freedom to repurpose embodiments of protected materials sold under the authority of the rights holder.[28] Some states permit private uses of protected material and research use exemptions.[29] Some have explicit provisions to facilitate interoperability and other transformational uses.[30] Patent laws also often include exceptions for experimentation, non-commercial use, and data generation for regulatory purposes.[31] And some states alter remedies to further protect the public interest.[32]

As Justice Brennan of the US Supreme Court said in a copyright fair-use case, *Harper & Row Publishers, Inc. v. Nation Enterprise,* at 589–90:

[Limitations are] not some unforeseen byproduct of a statutory scheme intended primarily to ensure a return for works of the imagination ... The copyright laws serve as the "engine of free expression," only when the statutory monopoly does not choke off multifarious indirect uses and consequent broad dissemination of information and ideas. To ensure the progress of arts and sciences and the integrity of First Amendment values, ideas and information must not be freighted with claims of proprietary right.

Similarly, human rights norms have informed the decisions of courts in the United Kingdom on the contours of British intellectual property law (Bently and Sherman 2009, 202; *Ashdown v. Telegraph Group Ltd.*).

These measures suggest that even when works are protected, there are important social uses that cannot be in the control of rights holders. Rights must be framed in ways that permit others to live a fully creative life – to build on existing works, play and learn from them, individuate them, and use them to satisfy intellectual inclinations. Intellectual property rights are not, in short, hermetically sealed. Rather, other values put pressure on the boundaries of each regime. The structure of some of these measures is also telling: many are open-ended and can accommodate unforeseeable social and technological changes.[33]

Controlling Abuse

Both intellectual property law and competition law are alert to the possibility that even with these limits, intellectual property rights can be used abusively. Thus, the Paris Convention allows states to issue compulsory patent licences to check abuse, including the failure to work (Article 5). The TRIPS Agreement similarly permits compulsory patent licences (Article 31). In addition, it recognizes that intellectual property rights can be misused and it permits states to regulate anticompetitive practices in all forms of intellectual property through cognate bodies of law (Articles 8 and 40).

States have, in fact, acted to protect the competitive environment. They treat intellectual property law and competition law as two sides of a coin and achieve balance through exceptions to intellectual property, through competition law, or through a combination of the two (Dreier 2001, 316). For example, the United States has recognized a misuse defence that would render patents unenforceable until the misuse is purged.[34] At the same time, it once recognized a strong antitrust cause of action. The latter action required an extensive factual basis, but cases were often brought by the government;

when private parties sued, they could obtain treble damages.[35] Together, misuse and antitrust law protected both the public interest and the rights of competitors. Reliance on these provisions has had a checkered career in the United States, but other countries routinely recognize that intellectual property rights can do more than encourage innovation. Because they can also be deployed to protect incumbents from innovation, a variety of tools are needed to safeguard the competitive and creative environment (Hemphill 2009; Hovenkamp, Janis, and Lemley 2006).

Applying the *Acquis* to Agriculture

Plant-specific intellectual property laws use many of the strategies described above. The UPOV Convention, for example, includes thresholds on protection (Articles 5–9) and specific limits on the scope of intellectual property, such as for acts done privately, for experimentation, and for certain types of breeding and growing (Article 15). The UPOV Convention recognizes the exhaustion principle and includes a public interest exception (Articles 16–17), and the duration of protection requirements are crafted to permit members to tailor their laws to the needs of specific fields.[36]

The same cannot be said when more general intellectual property laws are deployed to create exclusive rights in agricultural materials. For example, national utility patent laws were for the most part drafted before it became clear that plants would be entitled to utility patent protection, and before patent protection was sought for discoveries in biotechnology that are so fundamental that they cover agricultural innovation. Relative to other forms of intellectual property, there are also few domestic cases authoritatively interpreting national laws pertaining to plants.[37] Furthermore, although the WTO Dispute Settlement Board has signalled the importance of geographical indications relative to trademark rights (*Panel Report, European Communities – Protection of Trademarks and Geographical Indications for Agricultural Products and Foodstuffs*), it has yet to resolve a dispute involving any aspect of TRIPS that directly affects nutrition or agricultural research.

The *acquis* should therefore prove especially helpful in this realm. Consider, for example, rights over genes, proteins, and metabolic pathways, all of which constitute new ways to protect agricultural innovations, and one of which (gene patents) was the subject of the *Myriad* case discussed earlier. That case led to the invalidation of patents on genomic DNA, and the *acquis* explains the intuition that these patents might be invalid. Because the sequences encode principles of nature, they cannot be invented around. Accordingly, they prevent anyone who lacks permission from the rights holder from engaging in acts the *acquis* deems essential. The *acquis* offers three choices: the patents should be considered invalid, their scope should be delineated to facilitate crucial public uses, or refusals to license these uses should be regarded as antitrust violations.

Armed with that understanding, lawmakers, negotiators, and adjudicators should be better positioned to construct both domestic and international laws to deal with agriculture. For example, it is possible that the biotech industry will ask the US Congress to overturn the result in *Myriad* and enable gene patenting. If Congress were to take up that question, the *acquis* would suggest that it also pay attention to appropriate scope limitations. Or courts, even if opposed to broadening antitrust (competition) law generally, might be encouraged to draw an analogy between the holders of rights in subject matter like genes, which cannot be invented around, and the position of those who privatize essential facilities that could be duplicated (albeit at high costs). The same can be said at the international level. Consider, for example, the research exemption. Not without controversy, the United States has significantly restricted use of the research exemption in its own law. But any effort to impose these limits on other nations clearly falls far outside the agreed norms of intellectual property protection.

The *acquis* is also instructive on other issues arising in agriculture. For example, bag tag licences are typically framed as Stewardship Agreements, and some of the provisions may be needed to protect the environment.[38] But restrictions on research fall directly within the area where the *acquis* exposes a general consensus in favour of unrestricted access. Such measures should therefore be regarded as misuse, as anticompetitive, or as inconsistent with the first sale (exhaustion) doctrine. Conversely, and with the help of the rights holder portion of the *acquis* (Dinwoodie and Dreyfuss 2012, 189–92), it is clear that restrictions can be placed on the use of second-generation seed without violating the exhaustion doctrine because they would pertain to the alienation of material that the patent holder never sold or profited from.[39] At the same time, however, international agreements should give states the leeway to require rights holders to permit farmers to tinker with seed and customize successive generations to local needs. Health and nutrition are, after all, fundamental human rights (*Universal Declaration of Human Rights*, Article 25), well recognized in both national and international law as cabining the reach of intellectual property rights.

The brave new world of agricultural innovation has given rise to a brave new world of intellectual property protection for agricultural products and processes. Coming at a time when international intellectual property laws are expanding, it is particularly difficult for states to construct law that meets the demands of the industry while at the same time protecting the interests of growers, researchers, the new generations of breeders, and consumers. Much can be learned from examining existing intellectual property measures. When viewed holistically, it becomes apparent that the system overall encompasses a set of norms on what states, rights holders, and the public can legitimately expect. Professor Dinwoodie and I have attempted to begin an analysis of this implicit structure, but what we said in our book is not meant to be the final word; we hope other scholars will follow in our footsteps.

Agricultural innovation offers a particularly fertile ground for exploration. Because nutrition is undeniably a fundamental human right, access interests are especially acute. At the same time, research and development is risky and expensive. This is an area where multiple inputs are necessary for advancement and yet rights are often spread among multiple innovators. The need to regulate the food supply and the environment is clear, but intellectual property protection extends to the data that regulators require. This volume, which clarifies the many important issues at the intersection of agricultural science, intellectual property, and regulation, makes an important contribution to the literature and provides another window into the contents of the *acquis*.

Notes

1 See, e.g., *Ex Parte Hibberd*, and *J.E.M. Ag Supply, Inc. v. Pioneer Hi-Bred Intern., Inc.*, confirming the US Patent and Trademark Office's practice of awarding utility patents to plants.
2 The Seventh Circuit Court of Appeals enforced a shrinkwrap licence.
3 See, e.g., EC, *Council Directive 96/9/EC* (Database Directive).
4 Although developed, terminator seeds have never been approved for use.
5 See also 35 USC § 161.
6 See, e.g., 7 USC §§ 2402, 2541, 2453, 2544.
7 See 7 USC § 2404; see generally Janis and Kesan 2002.
8 The Supreme Court of the United States describes attempts to include compulsory licensing provisions in US patent law.
9 *Madey v. Duke Univ.* significantly limited the reach of *Whittemore v. Cutter*, in which Justice Story stated: "It could never have been the intention of the legislature to punish a man, who constructed such a machine merely for philosophical experiments,

or for the purpose of ascertaining the sufficiency of the machine to produce its described effects."

10 See, e.g., European Community 2005.

11 See also, e.g., *University of California v. Eli Lilly & Co.* and *Ariad Pharmaceuticals, Inc. v. Eli Lilly and Co.*

12 The declaratory plaintiffs have lodged an appeal; see OSGATA 2012.

13 For Germany, see *Gesetz zur Umsetzung der Richtlinie über den rechtlichen Schutz biotechnologischer Erfindungen* (Statute Implementing the European Council's Biotechnology Directive). For France, see *Code de la Propriété Intellectuelle.*

14 This case joined Cases C-241/91P and C-242/91P before the Court of Justice of the European Union.

15 For the activities of the WIPO Intergovernmental Committee on Intellectual Property and Genetic Resources, Traditional Knowledge and Folklore, see its website, http://www.wipo.int/tk/en/igc/index.html.

16 See *Trans-Pacific Partnership.* For other proposals, see http://www.citizen.org/leaked-trade-negotiation-documents-and-analysis.

17 See also Dinwoodie and Dreyfuss 2012, 147–56, noting the roles of the Food and Agriculture Organization, the World Health Organization, and the United Nations Conference on Trade and Development, among others.

18 See, e.g., *Anheuser-Busch v. Budejovicky Budvar NP* (U.K. C.A. 1984); *Anheuser-Busch Inc. v. Portugal* (Grand Chamber E.C.H.R. 2007); *Anheuser-Busch Inc. v. Budejovicky Budvar* (E.C.J. 2004); *Bud ě jovick ý Budvar v. OHIM* (C.F.I. 2008); *Budejovick ý Budvar v. Anheuser-Busch, Inc.* (C.J.E.U. 2011), arriving at conflicting decisions over the relationship between the Budweiser mark and geographical indication.

19 A similar view is apparent in *Golan v. Holder* (892), where the US Supreme Court dismissed the notion that there is an affirmative right to access: "Rights typically vest at the outset of copyright protection, in an author or rightsholder ... Once the term of protection ends, the works do not revest in any rightsholder. Instead, the works simply lapse into the public domain."

20 See also *Panel Report, United States – Section 110(5) of the U.S. Copyright Act,* para. 6.42, discussing the Berne *acquis.*

21 See, e.g., American Law Institute 1987, s. 702, and *Vienna Convention on the Law of Treaties,* Article 53.

22 See also TRIPS Agreement, Articles 7–8.

23 See also *WIPO Copyright Treaty,* Article 2.

24 See also *SAS Institute Inc v. World Programming Ltd,* para. 204.

25 See, e.g., *The Patents (Amendment) Act,* 2005, No. 15, ch. II, § 3d, Acts of Parliament, 2005 (India); see also *America Invents Act,* s. 14, codified as amended at 35 U.S.C. 102 note, altering the outcome of obviousness analyses for tax strategies.

26 See, e.g., Berne Convention, Article 9(2), and TRIPS Agreement, Articles 13, 17, 26, and 30.

27 See, e.g., US Constitution, Article I, § 8, cl. 8; compare with *Eldred v. Ashcroft.*

28 See, e.g., 17 U.S.C. § 109(a); *Quanta Computer, Inc. v. LG Electronics, Inc.*

29 See, e.g., EC, *Council Directive 2001/29/EC* (EP Copyright Harmonisation Directive), Articles 5.2–5.3.

30 See, e.g., ibid., Article 5.1; EC, *Council Directive 91/250/EEC* (Software Directive), Articles 6 and 5.2; 17 U.S.C. § 117.
31 See, e.g., *Patent Law* (Belgium), Article 28, s. 1(b): "The rights conferred by the patent shall not extend to acts done for scientific purposes with or on the subject matter of the patented invention"; *Consolidate Patents Act* (Denmark), s. 3(3)(iii): "The exclusive right shall not extend to: ... acts done for experimental purposes relating to the subject-matter of the patented invention"; *Patents Act 2004* (UK), s. 60(5)(b), excluding an act if "it is done for experimental purposes relating to the subject-matter of the invention"; *Patents Act 1977* (UK), s. 60(5)(b); *Panel Report, Canada – Patent Protection of Pharmaceutical Products,* WT/DS114/R (March 17, 2000).
32 See e.g., *eBay Inc. v. MercExchange, L.L.C.*
33 See, e.g., 17 U.S.C. § 107; TRIPS Agreement, Articles 13, 17, 26, and 30.
34 See, e.g., *Morton Salt Co. v. G. S. Suppiger Co.; Vitamin Technologists, Inc. v. Wisconsin Alumni Research Foundation.*
35 See, e.g., *United States v. Singer Mfg. Co.*
36 See UPOV Convention, Article 19, extending the minimum period of protection for trees and vines by five years.
37 In the United States, the Supreme Court's recent jurisprudence extends to just two cases: in *J.E.M. Ag Supply, Inc. v. Pioneer Hi-Bred Intern., Inc.,* the Court affirmed the patentability of new varieties of hybrid and inbred corn under the *Patent Act,* and in *Asgrow Seed Co. v. Winterboer,* the Court interpreted the crop exemption of the *Plant Variety Protection Act.*
38 See, e.g., Monsanto 2010 (Monsanto Technology Stewardship Agreement), especially s. 4, para. 9: "Grower may not plant and may not transfer to others for planting any Seed that the Grower has produced containing patented Monsanto Technologies for crop breeding, research, or generation of herbicide registration data. Grower may not conduct research on Grower's crop produced from Seed other than to make agronomic comparisons and conduct yield testing for Grower's own use."
39 See, e.g., *Bowman v. Monsanto Co.*

References

Legislation, Agreements, and Treaties

Agreement on Trade-Related Aspects of Intellectual Property Rights, April 15, 1994, *Marrakesh Agreement Establishing the World Trade Organization,* Annex 1C, Legal Texts: The Results of the Uruguay Round of Multilateral Trade Negotiations 320 (1999), 1869 U.N.T.S. 299, 33 I.L.M. 1197 [TRIPS Agreement].

America Invents Act, Pub. L. 112-29, 125 Stat. 284 (codified as amended in scattered sections of 35 U.S.C.).

Anti-Counterfeiting Trade Agreement. October 1, 2011. Office of the United States Trade Representative, http://www.ustr.gov/acta.

Berne Convention for the Protection of Literary and Artistic Works, September 9, 1896, as revised at Paris on July 24, 1971, 25 U.S.T. 1341, 828 U.N.T.S. 221 [Berne Convention].

Code de la Propriété Intellectuelle, Art. L613–2-1 (France).

Consolidate Patents Act, Consolidate Act No. 91 of January 28, 2009 (Denmark).

Convention on Biological Diversity, June 5, 1992, 1760 U.N.T.S. 79, U.N. Doc. ST/
    DPI/1307, 31 I.L.M. 818 (entered into force December 29, 1993) [CBD].
Declaration on TRIPS and Public Health, WT/MIN(01)/DEC/2 (November 14, 2001).
Doha Ministerial Declaration, WT/MIN(01)/DEC/1 (November 14, 2001).
EC, *Council Directive 91/250/EEC of 14 May 1991 on the Legal Protection of
    Computer Programs,* 1991 O.J. L. 122/42 [Software Directive].
EC, *Council Directive 96/9/EC of 11 March 1996 on the Legal Protection of Databases,*
    1996 O.J. L. 77/20 [Database Directive].
EC, *Council Directive 2001/29/EC of 22 May 2001 on the Harmonisation of Certain
    Aspects of Copyright and Related Rights in the Information Society,* 2001 O.J. L.
    167/10 [EP Copyright Harmonisation Directive].
European Patent Convention, October 5, 1973, 1065 U.N.T.S. 199 (as amended on
    December 13, 2007).
*Gesetz zur Umsetzung der Richtlinie über den rechtlichen Schutz biotechnologischer
    Erfindungen* (Statute Implementing the European Council's Biotechnology Dir-
    ective), January 21, 2005, BGBl I at 146, § 1a (4) (F.R.G.) (Germany).
International Convention for the Protection of New Varieties of Plants, December 2,
    1961, 815 U.N.T.S. 89 (as revised at Geneva on November 10, 1972, on October
    23, 1978, and on March 19, 1991) [UPOV Convention].
*Nagoya Protocol on Access to Genetic Resources and the Fair and Equitable Sharing
    of Benefits Arising from Their Utilization to the Convention on Biological Diver-
    sity, Conference of the Parties to the Convention on Biological Diversity,* UNEP/
    CBD/COP/DEC/X/1 of October 29, 2010. [Nagoya Protocol].
Paris Convention for the Protection of Industrial Property, March 20, 1883, as revised
    at Stockholm on July 14, 1967, 21 U.S.T. 1583, 828 U.N.T.S. 305 [Paris Convention].
Patent Law of March 28, 1984 (Belgium).
Patents Act 1977 (UK), 1977, c. 37.
Patents Act 2004 (UK), 2004, c. 16.
The Patents (Amendment) Act (India), 2005, No. 15, ch. II, § 3d, Acts of Parliament.
Plant Patent Act, Pub. L. No. 245, 46 Stat. 376 (codified as amended at 35 U.S.C.
    §§ 161–64 [2006]).
Plant Variety Protection Act, Pub. L. No. 577, 84 Stat. 1542 (codified at 7 U.S.C.
    §§ 2321–2582), see §§ 2321–72, 2401–2504, 2531–83.
Trade-mark Act, Act of March 3, 1881, ch. 138, 21 Stat. 502 (repealed 1946).
Trans-Pacific Partnership, Intellectual Property Rights Chapter, US Proposal, Draft
    – February 10, 2011. Knowledge Ecology International, http://keionline.org/sites/
    default/files/tpp-10feb2011-us-text-ipr-chapter.pdf [US Proposal].
United States Code, Title 17 (Copyright): 17 U.S.C. §§ 107, 109(a), 117.
United States–Chile Free Trade Agreement, June 6, 2003, 117 US Stat. 909.
United States–Morocco Free Trade Agreement, June 15, 2004, 118 US Stat. 1103.
Universal Declaration of Human Rights, December 8, 1948, G.A. Res. 217A (III),
    U.N. Doc. A/810.
Vienna Convention on the Law of Treaties, May 23, 1969, 1155 U.N.T.S. 331, 8 I.L.M.
    679.
WIPO Copyright Treaty, December 20, 1996, 2186 U.N.T.S. 121, 36 I.L.M. 65.

Cases

Anheuser-Busch Inc. v. Budejovicky Budvar, 2004 E.C.R. I-10989 (E.C.J. 2004).

Anheuser-Busch v. Budejovicky Budvar NP, [1984] F.S.R. 413 (U.K. C.A.).

Anheuser-Busch Inc. v. Portugal, App. No. 73049/01, 45 Eur. H.R. Rep. 36 [830] (Grand Chamber E.C.H.R., 2007).

Ariad Pharmaceuticals, Inc. v. Eli Lilly and Co., 598 F. 3d 1336, 1341 (Fed. Cir. 2010) (en banc).

Asgrow Seed Co. v. Winterboer, 513 U.S. 179 (1995).

Ashdown v. Telegraph Group Ltd., [2001] 3 W.L.R. 1368 (C.A.).

Association for Molecular Pathology v. Myriad Genetics, Inc., 132 S. Ct. 1794 (2012).

Association for Molecular Pathology v. USPTO, 653 F.3d 1329 (2011), judgment vacated, and remanded *sub nom. Ass'n for Molecular Pathology v. Myriad Genetics, Inc.,* 132 S. Ct. 1794 (2012).

Bilski v. Kappos, 130 S. Ct. 3218 (2010).

Bowman v. Monsanto, 133 S. Ct. 1761 (2013).

Bud ě jovick ý Budvar v. OHIM, [2009] E.T.M.R. 29 (C.F.I. 2008).

Budejovick ý Budvar v. Anheuser-Busch, Inc., [2012] E.T.M.R. 2 (C.J.E.U. 2011).

CCH Canadian v. Law Society of Upper Canada, [2004] 1 S.C.R. 339.

Dawson Chemical Co. v. Rohm and Haas Co., 448 U.S. 176, 215 (1980).

Diamond v. Chakrabarty, 447 U.S. 303, 315 (1980).

eBay Inc. v. MercExchange, L.L.C., 547 U.S. 388 (S. Ct., 2006).

Eldred v. Ashcroft, 537 U.S. 186 (2003).

Ex Parte Hibberd, 227 U.S.P.Q. 443 (C.C.P.A. 1985).

Golan v. Holder, 132 S. Ct. 873, 892 (2012).

Harper & Row Publishers, Inc. v. Nation Enterprise, 471 U.S. 539 (1985).

Harvard College v. Canada (Commissioner of Patents), 2002 SCC 76, [2002] 4 S.C.R. 45.

In re Fisher, 421 F.3d 1365 (Fed. Cir. 2005).

J.E.M. Ag Supply, Inc. v. Pioneer Hi-Bred Intern., Inc., 534 U.S. 124 (2001).

Madey v. Duke Univ., 307 F.3d 1351 (Fed. Cir. 2002).

Mayo Collaborative Serv. v. Prometheus Laboratories, Inc., 132 S. Ct. 1289, 1294 (2012).

Monsanto Canada Inc. v. Schmeiser, 2004 SCC 34, [2004] 1 S.C.R. 902.

Monsanto Tech. LLC v. Cefetra BV, 2010 E.C.R. 7, Case C-428/08.

Morton Salt Co. v. G. S. Suppiger Co., 314 U.S. 488 (1942).

Organic Seed Growers and Trade Ass'n v. Monsanto Co., 718 F.3d 1350 (Fed. Cir. 2013).

Panel Report, Canada – Patent Protection of Pharmaceutical Products, WT/DS114/R (March 17, 2000).

Panel Report, European Communities – Protection of Trademarks and Geographical Indications for Agricultural Products and Foodstuffs, WT/DS174/R (March 15, 2005).

Panel Report, United States – Section 110(5) of the U.S. Copyright Act, WT/DS160/R (June 15, 2000).

ProCD v. Zeidenberg, 86 F.3d 1447 (7th Cir. 1996).

Quanta Computer, Inc. v. LG Electronics, Inc., 553 U.S. 617 (2008).

Radio Telefís Eireann & Indep. Television Publ'ns Ltd. v. Comm'n of the European Communities, 1995 E.C.R. I-743, Joined Cases C-241/91 P & C-242/91 P.

SAS Institute Inc v. World Programming Ltd., [2010] EWHC 1829 (Ch).

United States v. Singer Mfg. Co., 374 U.S. 174 (1963).

University of California v. Eli Lilly & Co., 119 F.3d 1559 (Fed. Cir. 1997).

Verizon Comm'ns Inc. v. Law Offices of Curtis V. Trinko, LLP, 540 U.S. 398 (2004).

Vitamin Technologists, Inc. v. Wisconsin Alumni Research Foundation, 146 F.2d 941 (9th Cir. 1944).

Whittemore v. Cutter, 29 F. Cas. 1120, 1121 (C.C.D. Mass. 1813).

Secondary Sources

American Law Institute. 1987. *Restatement (Third) of the Foreign Relations of the United States.* Philadelphia: American Law Institute.

Bently, Lionel, and Brad Sherman. 2009. *Intellectual Property Law.* 3rd ed. Oxford: Oxford University Press.

Bently, Lionel, Brad Sherman, Denis Borges Barbosa, et al. 2010. "Experts' Study on Exclusions from Patentable Subject Matter and Exceptions and Limitations to the Rights." WIPO Standing Committee on Patents SCP/15/3 (September 2).

Benvenisti, Eyal, and George W. Downs. 2012. "The Democratizing Effects of Transjudical Coordination." *Utrecht Law Review* 8 (2): 158–71.

Broude, Tomer, and Yuval Shany. 2011. "The International Law and Policy of Multi-Sourced Equivalent Norms." In *Multi-Sourced Equivalent Norms in International Law,* ed. Tomer Broude and Yuval Shany, 1–17. Oxford: Hart Publishing.

Bugos, Glenn E., and Daniel J. Kevles. 1992. "Plants as Intellectual Property: American Practice, Law, and Policy in a World Context." *Osiris (2nd series)* 7: 75–104.

Carstensen, Peter. 2006. "Post-Sale Restraints via Patent Licensing: A 'Seedcentric' Perspective." *Fordham Intellectual Property Media and Entertainment Law Journal* 16 (4): 1053–80.

Dinwoodie, Graeme B., and Rochelle C. Dreyfuss. 2012. *A Neofederalist Vision of TRIPS: The Resilience of the International Intellectual Property Regime.* New York: Oxford University Press. http://dx.doi.org/10.1093/acprof:oso/978019530 4619.001.0001.

Dreier, Thomas. 2001. "Balancing Proprietary and Public Domain Interests: Inside or Outside of Proprietary Rights?" In *Expanding the Boundaries of Intellectual Property: Innovation Policy for the Knowledge Society,* ed. Rochelle Dreyfuss, Diane L. Zimmerman, and Harry First. New York: Oxford University Press.

Eisenberg, Rebecca S. 2001. "Bargaining over the Transfer of Proprietary Research Tools: Is the Market Failing or Emerging?" In *Expanding the Boundaries of Intellectual Property: Innovation Policy for the Knowledge Society,* ed. Rochelle Dreyfuss, Diane L. Zimmerman, and Harry First. New York: Oxford University Press.

European Community. 2005. *Submission of the European Community on the Convention of Biological Diversity, Global Status and Trends in Intellectual Property Claims: Genomics, Proteomics and Biotechnology.* UNEP/CBD/WG -ABS/3/INF/4 (January 11).

Foreign Affairs, Trade and Development Canada. 2013. "Trans-Pacific Partnership Free Trade Agreement Negotiations." August. http://www.international.gc.ca/trade-agreements-accords-commerciaux/agr-acc/tpp-ptp/index.aspx.

Gadbaw, R. Michael. 2010. "Systemic Regulation of Global Trade and Finance: A Tale of Two Systems." *Journal of International Economic Law* 13 (3): 551–74. http://dx.doi.org/10.1093/jiel/jgq031.

Helfer, Laurence. 2004. *Intellectual Property Rights in Plant Varieties: International Legal Regimes and Policy Options for National Governments.* FAO Legislative Study 85. Rome: Food and Agriculture Organization of the United Nations.

Hemphill, C. Scott. 2009. "An Aggregate Approach to Antitrust: Using New Data And Rulemaking to Preserve Drug Competition." *Columbia Law Review* 109 (4): 629–88.

Hovenkamp, Herbert, Mark D. Janis, and Mark A. Lemley. 2006. "Unilateral Refusals to License." *Journal of Competition Law and Economics* 2 (1): 1–42. http://dx.doi.org/10.1093/joclec/nhl002.

Huang, Kenneth G., and Fiona E. Murray. 2009. "Does Patent Strategy Shape the Long-Run Supply of Public Knowledge? Evidence from Human Genetics." *Academy of Management Journal* 52 (6): 1193–1221. http://dx.doi.org/10.5465/AMJ.2009.47084665.

Infojustice. 2012. "U.S. Trade Representative Ron Kirk Response to TPP Transparency Letter from Law Professors." Infojustice, http://infojustice.org/archives/26312.

Janis, Mark D., and Jay P. Kesan. 2002. "U.S. Plant Variety Protection: Sound and Fury ...?" *Houston Law Review* 39 (3): 727–78.

Juarez, Geraldine. 2012. "NAFTA Statement on ACTA and TPP." InfoJustice, http://infojustice.org/archives/9324.

Kapczynski, Amy. 2008. "The Access to Knowledge Mobilization and the New Politics of Intellectual Property." *Yale Law Journal* 117 (5): 804–85. http://dx.doi.org/10.2307/20455812.

Kaplan, Benjamin. 1967. *An Unhurried View of Copyright.* New York: Columbia University Press.

Kingsbury, Benedict, Nico Krisch, and Richard B. Stewart. 2005. "The Emergence of Global Administrative Law." *Law and Contemporary Problems* 68 (3–4): 15–61.

Louwaars, Niels, Hans Dons, Geertrui Van Overwalle, et al. 2009. "Breeding Business: The Future of Plant Breeding in the Light of Developments in Patent Rights and Plant Breeder's Rights." CGN Report 2009–14. Wageningen: Center for Genetic Resources, the Netherlands.

Mechlem, Kerstin. 2010. "Agricultural Biotechnologies, Transgenic Crops and the Poor: Opportunities and Challenges." *Human Rights Law Review* 10 (4): 749–64. http://dx.doi.org/10.1093/hrlr/ngq035.

Monsanto. 2010. "Monsanto Technology Stewardship Agreement." Farm Wars, http://farmwars.info/wp-content/uploads/2011/03/2010-Monsanto-Technology-Stewardship-Agreement-Downloadable-version.pdf.

Muscati, Sina. 2005. "Terminator Technology: Protection of Patents or a Threat to the Patent System?" *Idea* 45 (4): 477–510.

Nuffield Council on Bioethics. 2002. "The Ethics of Patenting DNA." Nuffield Council on Bioethics, http://www.nuffieldbioethics.org/patenting-dna.

O'Brien, Kevin J. 2012. "German Courts at Epicenter of Global Patent Battles among Tech Rivals." *New York Times,* April 8, 2012. http://www.nytimes.com/2012/04/09/technology/09iht-patent09.html?pagewanted=all.

O'Rourke, Maureen A. 2000. "Toward a Doctrine of Fair Use in Patent Law." *Columbia Law Review* 100 (5): 1177–1250. http://dx.doi.org/10.2307/1123488.

OSGATA (Organic Seed Growers and Trade Association). 2012. "Battle over Farmers Rights against Monsanto Continues to Brew." OSGATA, http://www.osgata.org/farmers-determined-to-defend-right-to-grow-food-file-appeal-in-osgata-vs-monsanto.

Rowe, Elizabeth A. 2011. "Patents, Genetically Modified Foods, and IP Overreaching." *SMU Law Review* 64 (3): 859–93.

Slade, Alison. 2011. "Articles 7 and 8 of the TRIPS Agreement: A Force for Convergence within the International IP System." *Journal of World Intellectual Property* 14 (6): 413–40.

Strandburg, Katherine J. 2011. "Patent Fair Use 2.0." *UC Irvine Law Review* 1: 265–306. Also available at Social Science Research Network, http://papers.ssrn.com/sol3/papers.cfm?abstract_id=1835007.

Strauss, Debra M. 2009. "The Application of TRIPS to GMOS: International intellectual Property Rights and Biotechnology." *Stanford Journal of International Law* 45 (2): 287–320.

TPP Watch. 2012. "TPP Watch Bulletin #20."October 29. It's Our Future, http://www.itsourfuture.org.nz/tpp-watch-bulletin-20/.

Yu, Peter K. 2011. "Six Secret (and Now Open) Fears of ACTA." *SMU Law Review* 64 (3): 975–1094.

Contributors

VERNON BACHOR is an associate professor of entrepreneurship and innovation at Winona State University, a business consultant on research-based technology and innovation, and the managing editor of the *Journal of Engineering and Technology Management*. Vern's research is an interdisciplinary approach to understanding the interrelations of strategy, technology and innovation management, sustainability, technical knowledge diffusion, entrepreneurship, international business, project/program management, and MIS. He has most recently published in *California Management Review*, *Technovation*, and *Research-Technology Management*.

HANNES DEMPEWOLF is a member of the Science Team at the Global Crop Diversity Trust and manages the project "Adapting agriculture to climate change: collecting, protecting and preparing crop wild relatives." Hannes has a particular interest in exploring ways in which crop wild relatives and land race diversity can be used in breeding programs more effectively by better linking gene banks to breeders (and vice versa). His scientific interests focus on the evolution, maintenance and conservation of agro-biodiversity, the importance of such diversity for farming communities, and the role it can play for sustainable development and food security.

ROCHELLE COOPER DREYFUSS is the Pauline Newman Professor of Law at New York University School of Law and co-director of its Engelberg

Center on Innovation Law and Policy. She is a member of the American Law Institute and was a co-reporter for its Project on Intellectual Property: Principles Governing Jurisdiction, Choice of Law, and Judgments in Transnational Disputes. She was a consultant to the Federal Courts Study Committee, to the Presidential Commission on Catastrophic Nuclear Accidents, to the Federal Trade Commission, and she served on the Secretary of Health and Human Services' Advisory Committee on Genetics, Health, and Society.

REGIANE GARCIA is a foreign-trained lawyer from Brazil, where she practised at all levels of court and various administrative tribunals and agencies. She is currently a PhD candidate at the University of British Columbia Faculty of Law. Her current area lies at the intersection between law, population health, and governance of health systems. In particular, the focus of her work is on legal frameworks and deliberative structures that govern and operationalize collaborations between state and civil society at large in the design, implementation, and evaluation of public health policies.

R. NELSON GODFREY has written extensively on issues relating to intellectual property and regulatory regimes and their interaction with research and knowledge translation, with particular emphasis on topics relating to genomics research and living modified organisms. Nelson is also an associate in the Vancouver offices of a leading Canadian intellectual property and technology law firm whose practice includes advising clients on litigation, compliance and regulatory matters, and licensing transactions.

GREGORY GRAFF is an associate professor in Agricultural and Resource Economics at Colorado State University. He specializes in the economics and public policy of technological innovation and technology-based entrepreneurship and how they can drive sustainable economic development. His areas of expertise include innovation in agricultural biotechnology and crop genetics; intellectual property in biotechnology; the role of regulatory approvals as gatekeeper of new technologies; how economic interests drive the politics of regulation; and the policy, economics, and management of technology transfer.

JEREMY HALL is a professor at the Beedie School of Business, Simon Fraser University (on leave), chaired professor of corporate social responsibility/

sustainable business, and director of the International Centre for Corporate Social Responsibility at Nottingham University Business School. He is also editor-in-chief of the *Journal of Engineering and Technology Management*, a technology and innovation management journal with a 2013 impact factor of 2.101. His research interests include innovation dynamics, the social impacts of innovation and entrepreneurship, sustainable supply chains, and strategies for sustainable development innovation. Much of his research has been conducted in Brazil.

SARAH HARTLEY is a research fellow at the University of Nottingham. Sarah investigates the science/policy interface in the context of emerging biotechnologies in Europe and North America. She has published in several policy journals and produced an edited volume on science, ethics, and public policy. Previously, she held positions at Simon Fraser University, the University of British Columbia, the University of Victoria, and Genome British Columbia.

RONALD J. HERRING is Professor of Government and International Professor of Agriculture and Rural Development at Cornell University. Ronald has worked mostly in and on South Asia, in fields of agrarian political economy, ethnicity and conflict, political ecology and development, and social conflicts around science and genetic engineering. Ronald was editor of the new *Oxford Handbook of Food, Politics, and Society* (2015) and published "State Science, Risk and Agricultural Biotechnology: Bt Cotton to Bt Brinjal in India" (*Journal of Peasant Studies*, 2015). A current book project is tentatively entitled *Suicide Seeds and Silver Bullets: GMOs from the Ground Up*.

RACHAEL MANION is a senior associate at a public affairs consulting firm in Ottawa, Canada, advising clients in the health and food sectors. She draws on her work as a regulatory lawyer for the federal Department of Justice, where she advised Health Canada on the development and application of statutory regimes to novel technologies and the development of science policy.

EMILY MARDEN is a research associate at the University of British Columbia Faculty of Law, where she teaches and directs research on genomics and innovation in agriculture and the life sciences, and counsel to Sidley Austin LLP Food and Drug Regulatory practice in Palo Alto, California. She has published and presented widely on interrelated issues of intellectual property, regulation, and innovation in the agricultural and life sciences arenas, and she continues to work toward enabling innovation in agricultural genomics through research and her involvement with international organizations such as the FAO's Treaty Secretariat and DivSeek, a multistakeholder organization dedicated to bridging the gap between gene banks, academic researchers, and breeders.

STELVIA MATOS is an adjunct professor at the Beedie School of Business, Simon Fraser University. She has published in *Energy Policy, Journal of Management Studies, Journal of Operations Management, Journal of Cleaner Production, International Journal of Production Research, Technovation, Research Policy, Harvard Business Review (L.A. Edition)*, and the *Journal of Business Ethics*. Her research areas include technological, commercial, organizational and social uncertainties of innovation, life-cycle assessment, technical-economic cost modeling, sustainable development innovation, and social aspects of innovation dynamics. Her research involves the agriculture, aquaculture, chemical, energy, forestry and tourism sectors, with field studies performed in Brazil, Bosnia, Canada, China, Italy, the Netherlands, the UK, and the US.

CHIDI OGUAMANAM is a full professor in the Faculty of Law at University of Ottawa (Common Law Section). He is affiliated with the Centre for Law, Technology and Society, the Centre for Environmental Law and Global Sustainability and the Centre for Health Law, Ethics and Policy at the University of Ottawa. He writes, teaches, and consults in the areas of law and technology (including biotechnologies), agricultural knowledge systems, food security and global food systems, Indigenous peoples and Indigenous knowledge, intellectual property, human rights and global knowledge governance, and contract law. He is a fellow of Canada Institutes of Health Research Program in Ethics of Health Research and Policy, and a Senior Research Fellow in the Centre for International Sustainable Development Law at McGill University.

MATTHEW R. VOELL is a lawyer practising at Underhill Gage Litigation. Matthew has published a number of articles on the intersection of law and genomics. His current areas of focus include environmental, administrative, aboriginal and constitutional law, and he maintains a broad general civil litigation practice. Matthew has appeared before all levels of court in British

Columbia, the Federal Court of Canada, and numerous administrative tribunals and agencies.

DAVID ZILBERMAN is a professor and holds the Robinson Chair in the Department of Agricultural and Resource Economics at U.C. Berkeley. David's areas of expertise include agricultural and environmental policy, the economics of innovation, risk and marketing, water, pest control, biotechnology, and climate change. David is a fellow of the American Agricultural Economics Association (AAEA) and the Association of Environmental and Resource economics (AERE), and is the recipient of the 2000 Cannes Water and the Economy Award. He won the AAEA 2002 and 2007 Quality of Research Discovery Award and the 2005 and 2009 AAEA Publication of Enduring Quality Award.

Index

Aber, Josh, 186

Acs, Zoltan J., 175

African Agricultural Technology Foundation, 93

Agreement on Trade-Related Aspects of Intellectual Property Rights. See TRIPS Agreement (*Agreement on Trade-Related Aspects of Intellectual Property Rights*)

agricultural biotechnology: definition of, as used in book, 43; early views of risks of agricultural biotechnology, 44–45; expressed aim of, 180; genomics, proteomics, and cloning included, 43; linked with multinational capital (due to Monsanto in India), 104

agricultural biotechnology, social and ethical concerns: concerns labelled as "emotional" or "religious," 53; consideration recommended by official reports (late 1990s), 49–50; debate over socio-economic risks of GM wheat, 52–57; emergence in discussion re regulation, 42–43, 46– 47, 49–50; European concern, 46, 49, 57; issues of concern, 42–43, 46– 47, 52; not addressed in regulatory framework, 43–46, 47–49, 57–59, 60-62; public confidence needed, 50; Royal Society of Canada's Expert Panel report (2001), 51–52; scientism and, 44, 58–59; socioeconomic risks to be considered (2003), 53

agricultural genomics: academic coverage of IP–Regulatory Complex's impact on, 16; policy development and ITPGRFA, x; role of IP–Regulatory Complex in, 4–6. *See also* agricultural biotechnology; agricultural biotechnology, social and ethical concerns; agricultural genomics research; agricultural innovations; GMOs (genetically modified organisms); *entries beginning with* intellectual property

agricultural genomics research: benefits of, 15, 70–71; benefits of genomics, according to proponents, 70–71, viii;

challenges in translating into socially
beneficial products, 3–4, 21;
Genome Canada, 15; impact of IP
and biosafety regimes, 4, 15–16;
invalidation of patents on genomic
DNA, 224–25; LMOs (living modified organisms), 20–21, 34n1, 35n4–
5; needed to address global hunger
and poverty, 161–62; patents (*see*
patents); rDNA discovery and private sector interest in agriculture:
focus on economically viable monocultures, 143; Responsible Research
and Innovation (RRI) concept, 61;
sunflower genomics project, 16, 18,
33, 37n18; in the US, 15. *See also*
agricultural innovations; GMOs
(genetically modified organisms)
agricultural innovations: *Bt* insect-
resistant crops (*see Bt* insect-resistant
crops); Canadian efforts to minimize
regulatory burdens, 19; drought-
tolerant crops, 18, 75, 82–83, 87;
gene use–restriction technologies
(GURT) only way to restrict gene
flow, 128; herbicide tolerance (*see*
Roundup Ready–tolerant crops);
hybridization and proprietary protection, 5, 10n3; IP rights relatively
recent (early 1900s), 5; legitimization of innovations (cognitive,
socio-political), 176–78, 186–87;
little academic attention on impact
of IP–Regulatory Complex, 16; patent efficacy for product *vs.* process
innovations, 173; plant "variety,"
characteristics defined (new, distinct, uniform, stable), 10n5; socially
beneficial (*see* socially beneficial
products). *See also* innovation
incentives, socially beneficial
products, and the IP–Regulatory
Complex
Agriculture and Agri-Food Canada:
member of Canadian Biotechnology
Advisory Committee, 50–51;

proponent of GM wheat, 54; public
"education" re GM food, 58
Aldrich, H., 176
Anti-Counterfeiting Trade Agreement
(US), 217
anticommons hypothesis, 76
Arbour, Louise, Justice, 31
Argentina and Roundup Ready soy, 118
Asgrow Argentina, 118
*Association for Molecular Pathology
v. Myriad Genetics, Inc.*, 79, 214–15,
224–25
Australia and patents on plants, 30

Barwale, B.R., 103
BASF Group, 84
Baumol, William, 174
Bayer, 84
Benefit Sharing Fund (of ITPGRFA),
150–52, 155, 163
Benvenisti, Eyal, 221
Berne Convention, 216–17, 221, 222
Bilski v. Kappos (US), 221–22
*Biopiracy: The Plunder of Nature and
Knowledge* (Shiva), 103
biosafety, 4
Biosafety Law (Brazil, 1994), 116
biosafety regimes: basic goal of bio-
safety, 77; Brazil's CTNBio federal
agency, 116–17; Canada's Plant
Biosafety Office, 27; CBD and, 5, 19,
20–21, 35n5; cost-benefit analysis of
Bt cotton in India, 126–28; domestic
frameworks inconsistent, 5; goals, 5;
impact on agricultural innovation,
4, 21; as impediments to product
approvals, 3; India's regulation based
on incremental risk, 127; role of
WTO agreements, 19–20. *See
entries beginning with* regulation
Biotechnology Industry Organization,
79
*Biotechnology Regulation in Canada:
A Matter of Public Confidence*
(1996), 49
Bioversity International, 25

Brackenridge, Peter, 56

Brazil: corporate globalization a target for opposition of transgenic soy, 117, 119; criticism of expert-regulatory model of policy-making, 198; CTNBio federal agency responsible for biosafety, 116–17; disjointed regulatory environment for trans-genics, 118; GMO-free soy threatened by stealth soy, 119; governments' differing support for transgenic soy, 117, 119–20, 183; Greenpeace's opposition to regularization of transgenic soy, 116–17, 183; inequal-ity among farmers (small- *vs.* large-scale), 182–83; Institute for Consumer Defense's opposition to regulariza-tion of of transgenic soy, 116–17, 183; Landless Rural Workers' Movement against GMO food, 118–19; laws on transgenics, IP and regulatory, 116; legitimacy to rule on transgenics contested, 116–17; Monsanto a target for opposition to transgenic soy, 115, 117, 182–83; Monsanto's rigid IP approach, impact on small- *vs.* large farmers, 182–83; official support for transgenics but challenges from civil society, 115–18, 120; opponents to regularization of of transgenic soy, 116–17, 183; property rights depend-ent on monitoring by technology, 123; public researcher EMBRAPA's new transgenic soy, 121, 182–83; soy a major export commodity, 115. *See also* stealth seeds in Brazil

Brazilian Enterprise for Agricultural Research (EMBRAPA), 121, 182-83

Brunk, Conrad, 51

Bt insect-resistant crops: activists in India burning *Bt* cotton trials, 103, 111, 112–13, 127; biosafety testing in India, 126–28; *Bt* microbe a targeted biopesticide, 71–72, 95n4; concerns/ protests by public, 72; Cry1Ac gene in, 105, 110, 113, 114; economic benefits in US, 72; farmers' adoption of both Monsanto and public sector versions of *Bt* cotton, 109–10; impact on stakeholders, 72–74; market exclusivity in India for Monsanto/ Mahyco despite no patent, 80–81; MECH-162 variety of *Bt* cotton with non-*Bt* hybrid, 110–11; official ap-proval of *Bt* cotton in India (2002), 107, 127; performance variation in *Bt* cotton and "failure" reports, 110, 113; resistance development in cotton not a problem, 126; social benefits of, 74; social welfare analysis of *Bt* corn, 72–74; stealth seeds in India, 104–9

Bugos, Glenn, 212

Caldwell, R.B., 46

Campaign for a Transgenic-Free Brazil, 120

Campolete, Solet, 118–19

Canada: champion of TRIPS Agreement, 146; *Food and Drugs Act* on safety of agricultural prod-ucts, 17; genomics research through Genome Canada, 15. *See also* intel-lectual property regime, Canadian; regulation of agricultural biotech-nology, Canadian

Canadian Agricultural Research Council, 46

Canadian Biotechnology Advisory Committee (CBAC): considered social/ethical concerns "emotional," 53; establishment (late 1990s) and mandate, 44, 50; limited under-standing of social/ethical issues, 51; public consultations criticized for bias (2000), 51; report and conclu-sions (2002), 51; termination (2007), 60

Canadian Environmental Network, 48

Canadian Environmental Protection Act (CEPA) (Canada, 1988), 47

Canadian Federation of Agriculture,
54, 56
Canadian Food Inspection Agency
(CFIA): "familiar" and "substantially
equivalent" criteria, 27, 36n13;
market acceptability as criterion
for regulatory approval, 56;
"novelty" of a variety trigger for
regulatory review, 27, 96n21;
Plant Biosafety Office, 26; primary
regulatory body re novel plants,
26; product rather than process
analyzed, 26–27; regulation based
solely on scientific safety of products, 60; on risks of agricultural
biotechnology, 45; social/ethical
concerns re biotechnology seen
as "religious," 53
Canadian Health Coalition, 58
Canadian Institute for Environmental
Law and Policy, 46–47
Canadian International Development
Agency (CIDA), 58
Canadian National Millers Association,
55–56
Canadian Seed Growers' Association,
54
Canadian Wheat Board, 54, 55–56,
62n4
*Capturing the Advantage: Agricultural
Biotechnology in the New Millennium*
(1998), 49–50
Cardoso, Fernando, 117
Cartagena Protocol on Biosafety:
regulation of genetically modified
plants, 20–21, 35n5
Casale, Carl, 57
CBAC. *See* Canadian Biotechnology
Advisory Committee (CBAC)
CBD. *See Convention on Biological
Diversity* (CBD)
CEPA. *See Canadian Environmental
Protection Act* (CEPA) (Canada,
1988)
CFIA. *See* Canadian Food Inspection
Agency (CFIA)

CGIAR. *See* Consultative Group for
International Agricultural Research
(CGIAR)
CGIAR Consortium: areas of interest,
155–56; farmers' rights treated with
equivocation, 160; formation (2008),
155; intellectual assets (IA or IP),
approach in Principles, 157–59;
IP rights denied only for the form
received by CGIAR, 159–60; judicial
weakness in treating IP, 162; jurisdictional overlaps with ITPGRFA, 160–
61; Principles and Guidelines on
the Management of Intellectual
Assets, 156–57. *See also International Treaty on Plant Genetic
Resources for Food and Agriculture*
(ITPGRFA)
Chataway, Joanna, 180
CIDA. *See* Canadian International
Development Agency (CIDA)
cognitive legitimization of innovations,
176, 178, 186–87
Comissao Tecnica Nacional de
Biosseguranca (CTNBio federal
agency [Brazil]), 116–17
Commission on Plant Genetic Resources
for Food and Agriculture (CPGRFA):
collaboration with ITPGRFA, 161;
Global Plan of Action (FAO), 149,
161, 166n5
Common Implementation Strategy (CIS)
of *European Union Water Frame-
work Directive*, 208
Consultative Group for International
Agricultural Research (CGIAR):
description, 152–53; HarvestPlus
(HP) program, 154; IARCs, 152–55,
158; reform (2008) and renamed
CGIAR Consortium, 155; relationship with ITPGRFA, 154–55; seed
banks held in trust for ITPGRFA
in IARCs, 145, 152–54. *See also
International Treaty on Plant
Genetic Resources for Food and
Agriculture* (ITPGRFA)

Consumer Defense Code (Brazil, 1990), 116

Convention on Biological Diversity (CBD): allows for wide variability among biosafety regimes, 5; biodiversity and access and benefit-sharing objectives same as ITPGRFA's, 147–48; biosafety and health measures for novel plants, 19; *Cartagena Protocol on Biosafety*, 20–21, 35n5; jurisdictional overlaps with ITPGRFA, 160–61; *Nagoya Protocol* (2014), 161, 164, 217; potential for inconsistent national regimes, 35n9

Council of Canadians, 54

CPGRFA. *See* Commission on Plant Genetic Resources for Food and Agriculture (CPGRFA)

CropLife Canada, 54

crops. *See* socially beneficial products

de Janvry, A., 122

Desai, Bipin, 114

Desai, D.B., 105

Diamond v. Chakrabarty (US), 35n8, 212–13

Dinwoodie, Graeme B., 212, 218–24

Dixit, A.K., 111

Dow Chemical Company, 103

Downs, George, 221

Dreyfuss, Rochelle C., 212, 218–24

drought-tolerant crops: commercialization lacking (drought tolerance an orphan trait), 75, 87; countries doing research in, 83; effect of IP-Regulatory Complex on development of, 82–83; research mainly for soybeans, 82–83; silverleaf sunflower, 18

Drucker, P., 178

Duke, L.H., 46

Dupont Pioneer (*formerly* Pioneer Hi-Bred), 83–84

Eisenberg, Rebecca, 215

Elements of Precaution: Recommendations for the Regulation of Food

Biotechnology in Canada (Royal Society of Canada), 51–52

Ellis, Brian, 51

EMBRAPA. *See* Brazilian Enterprise for Agricultural Research (EMBRAPA)

Environment Canada, 47

EU Framework Programme for Research and Innovation: Responsible Research and Innovation (RRI) concept, 61

Europe: approval of UPOV Convention, 22; body of principles in IP *acquis*, 220; champion of TRIPS Agreement, 146; gene patents limited to utility disclosed, 215–16; growing concern about social/ethical issues in agricultural biotechnology, 46, 49, 57; nontransgenic R&D methods due to costs, 87; plant breeders' rights under UPOV, 146; preference for GMO-free food, 116, 119, 181–82; regulatory approvals of GM products temporary, 81–82; Responsible Research and Innovation (RRI) concept, 61; socio-political concerns re Monsanto's right IP management policies, 181–82; TRIPS-plus agreements, 24

European Union Water Framework Directive, 9, 205–7, 208

Ex parte Hibberd (US), 30

FAO (Food and Agriculture Organization). *See* United Nations Food and Agriculture Organization (FAO)

farmers' rights/privilege (to save and replant seed): goal of ITPGRFA, 145, 148–49; inconsistency between *Patent Act* and *Plant Breeders' Rights Act*, 31; part of IP *acquis*, 225; under *Plant Breeders' Rights Act* (Canada), 29, 31; "privilege" under UPOV, 23, 31, 35n7, 160, 167n15; "rights" under CGIAR but treated with equivocation, 160; seed saving, Monsanto's position, 183–84;

seed saving permitted for most of
recent history, 5, 10n2
Federal Framework for Biotechnology
(Canada, 1993), 47
Feeds Act (Canada), 48
Fertilizers Act (Canada), 48
Fetterhoff, Terry, 186
Fiol, M., 176
Fischer, Frank, 198
Food and Agriculture Organization
(FAO). *See* United Nations Food and
Agriculture Organization (FAO)
France and gene patents, 215
Friends of the Earth, 58
Fung, Archon, 202, 203
Funtowicz, Silvio, 199
Furtan, W.H., 54

General Agreement on Tariffs and Trade
(GATT, WTO), 19–20
Genetic Engineering Approval Committee (GEAC) (India), 105–7, 109,
127
Genome Canada, 15. *See also* sunflower
genomics project (Canada)
genomics. *See* agricultural genomics
research
Germany and gene patents, 215
Gervais, Daniel, 211
Global Plan of Action (FAO), 149, 152,
161, 166n5
globalization and biotechnology:
activism in India against multinational capital, 102–4; corporate
globalization a target for opposition
of transgenic soy in Brazil, 117, 119;
international impact of national
policies, 78–79, 211–12, 216–18;
suicides by debt-ridden farmers
in India, 103
glyphosate. *See* Roundup Ready–
tolerant crops
GMOs (genetically modified organisms):
determining regulatory framework,
45; GM wheat, opponents and proponents, 54; GM wheat, regulation

and socio-economic issues, 52–53,
55–56; GM wheat's market acceptability and regulatory approval,
56–57; main issue in agricultural
biotechnology, 43; modified and
stealth soy in Brazil, 118–19; opposition to in agriculture, 3; regulation
scattered under various Acts, 48;
uniform conception lacking, 34n1
Godfrey, R. Nelson, 156
Gokhale, A.M., 109
governance and the IP–Regulatory
Complex: allocation of decision-
making responsibilities as part of
law, 218–19; call for participation
of non-state actors, 196, 199, 202,
203–4; challenges of exclusion of
stakeholders and secrecy, 195, 202;
definition of governance, 199, 200;
*European Union Water Framework
Directive* as example of new governance, 9, 205–7, 208; new configurations (complementarity, rivalry, and
transformation), 201–2; new governance model, 196; new institutional
arrangements and principles, 203–
4; new legal features, 199, 200–1;
new *vs.* traditional models, 200–2;
participatory governance model,
204–5, 207; public participation
absent in regulatory model, 195–96;
regulatory model, criticism of, 198–
99, 207; regulatory model, description, 197–98, 201–2; skepticism re
new model and rebuttals, 202–3;
socio-ethical concerns marginalized
in regulatory model, 195
Grain Growers of Canada, 53, 54
Gray, R.S., 54
Greenpeace: appeal against approval of
transgenic soy in Brazil, 116, 183; on
efforts to downplay risks of agricultural biotechnology, 58; opposition
to GM wheat, 54, 58

Habermas, Jürgen, 199, 203, 204

Hall, Jeremy, 176, 177–78
Hardy, Cynthia, 174
Harper & Row Publishers, Inc. v. Nation Enterprise (US), 222–23
Harries, Adelaida, 118
Harvard College v. Canada (2002), 30, 216
HarvestPlus (HP) program (CGIAR), 154
Health of Animals Act (Canada), 48
Herring, Ronald J., 183, 213
Holzman, J.J., 54
Horizon 2020, 61
House of Commons Standing Committee on Agriculture and Agri-Food: dismissal of socio-economic risks of GM wheat, 53, 55; on GM wheat issue, 56–57; report on biotechnology regulation (1998), 49–50
House of Commons Standing Committee on the Environment and Sustainable Development, 49

IARCs. *See* International Agricultural Research Centres (IARCs)
IDEC. *See* Institute for Consumer Defense (IDEC) (Brazil)
India: biosafety regime controlled by politicians and farmers, 106; biosafety regulation based on incremental risk, 127; *Bt* cotton (*see Bt* insect-resistant crops); cost-benefit analysis of *Bt* cotton, 126–28; Deshiya Karshaka Samrakshana Samithi (DKSS) farmer organization, 102; farmer organization (DKSS) to fight globalization, 102–3; farmer organization (KRRS) to fight transgenic crops, 111–12; Genetic Engineering Approval Committee, 105–7, 109, 127; *Indian Seeds Act, 1966*, 109; Industrial Toxicology Research Centre, 127; International Crops Research Institute for the Semi-Arid Tropics, 92; Khedut Samaj (farmer organization), 114;

Mahyco company, 80–81, 103–5, 107–8, 126; nationalist argument against transgenic crops, 113–14, 183; official approval of *Bt* cotton (2002), 107, 127; Platform for Translational Research on Transgenic Crops, 92; suicides by debt-ridden farmers, 103, 104; transgenic anti-bollworm crop in Gujarat, 104–6. *See also Bt* insect-resistant crops; stealth seeds in India
Indian Express story of stealth seeds in Gujarat, 108
Indian Seeds Act, 1966, 109
Industrial Property Code (Brazil, 1996), 116
Industrial Toxicology Research Centre (India), 127
innovation: definition, 174; encouraged and endangered by IP, 88–89, 174–75, 213, ix; outcome of "productive entrepreneurship," 174–75; Responsible Research and Innovation concept (Europe), 61. *See also* agricultural innovations; innovation incentives, socially beneficial products, and the IP–Regulatory Complex
innovation incentives, socially beneficial products, and the IP–Regulatory Complex: combined IP and regulatory incentives, 92–93, 97n29; complementary mechanisms of market exclusion, 79–82; effect on development of crops with improved nutrition, 84–85; effect on development of drought-tolerant crops, 82–83; impact of biosafety regimes for agricultural innovation, 4, 21; IP for genetic materials, arguments for and against, 75–76; IP incentives for, 88–89; IP management, suggestions for effective innovation, 185–87; IP and regulatory systems closely intertwined, 68–69; market exclusivity with no patent protection, 80–82; need to navigate both IP and

regulatory systems, domestic and foreign, 15–16, 78–79; regulatory systems and incentives, 76–78, 90–91
Institute for Consumer Defense (IDEC) (Brazil), 116, 183
intellectual property *acquis* for agriculture: allocation of decision-making responsibilities as part of law, 218–19; applying the *acquis*, 224–25; breadth of gene patents, 214–15; controlling abuse, 223–24; core principles with holistic view of IP, 212, 219–21, 225–26; distributive justice and increased IP protection, 213–15; helpful in constructing domestic and international laws, 224–25; international pressure on states to protect agriculture, 78–79, 211–12, 216–18; issue of farmers' rights, 225; issues re Stewardship Agreements, 225; nonlegal strategies to protect, 213; plant patents, 212–15; principles of explicit recognition of user interests, 220; principles as *jus cogens* of IP law, 221; rights and exceptions, differing interpretations, 219; scope limitations, 222–23; subject matter exclusions, 221–22; "terminator" technology, 213
intellectual property management: case study methodology, 178–80; in countries with weak institutions, 175, 178, 184–87; effective innovation, suggestions, 185–87; institutional context for IP protection, 174–75; IP as valuable asset needing protection, 171; legitimization of innovations (cognitive, socio-political), 176–78, 186–87; logic and limitations of protection, 172–74; Monsanto's IP approach and consequences, 180–84, 185, 187; patent efficacy for product *vs.* process innovations, 173; rigid approaches as problematic, 173, 180–83, 184, 186–87; "Schumpeterian rents," 176–77;

technological gatekeepers, 185–86. *See also* intellectual property protection
intellectual property protection: argument for (social contract), 79–82, 174–75, 213; arguments for and against IP in genetic materials, 75–76; Berne Convention, 216–17, 221, 222; definition, 4; distributive justice and increased IP protection, 213–15; institutional context for IP protection, 172–74; international pressure on states to protect agriculture, 78–79, 211–12, 216–18; IP as valuable asset needing protection, 171; IP rights in agriculture recent (early 1900s), 5, 10n1; moral arguments, 76; nonlegal strategies to protect, 172, 213; Paris Convention, 216–17, 223; patents (*see* patents); realist/political economy arguments, 76; reasons for *not* patenting, 174; social contract: IP protection *vs.* costly regulation, 79; "terminator" technology, 213; tightening patent criteria to maximize innovation, 88; as valuable asset needing protection, 171. *See also Monsanto Canada Inc. v. Schmeiser* (2004)
intellectual property regime, Canadian: Canada party to CBD and WTO regimes, 26; national policies based on national priorities, 26, 28; *Plant Breeders' Rights Act*, 29, 31–32, 36n16; Plants with Novel Traits (PNT), 4, 17, 26–27, 33, 37n18; *Seeds Act*, 26, 48, 56. *See also* intellectual property regimes, international; *Patent Act* (Canada); regulation of agricultural biotechnology, Canadian
intellectual property regimes, international: *acquis* for (*see* intellectual property *acquis* for agriculture); anticommons hypothesis, 76; barriers to the poor, 175; Berne

Convention, 216–17, 221, 222; breeders' rights enactments, 22–23; difficulty in pleasing industry, researchers, and consumers, 226; distributive justice and increased IP protection, 213–15; focus on economically viable monocultures, 143; gene use–restriction technologies (GURT) only way to restrict gene flow, 128; goals, 16; as impediments to developing products, 3; incentives and (*see* innovation incentives, socially beneficial products, and the IP–Regulatory Complex); move toward sharing of plant IP and genetic resources, 22, 24–25; national policies based on national priorities, 26; Paris Convention, 216–17, 223; pressure on states to protect agriculture, 78–79, 211–13, 216–18; protecting public access and future innovation, 212; rDNA discovery and private sector interest in agriculture, 143; recent developments, 21, 22–24; seed industry sought property rights, 21; social contract: IP protection *vs.* costly regulation, 79; standardized approaches, problems, 173; where institutions are weak, 175. *See also Convention on Biological Diversity* (CBD); intellectual property protection; *International Convention for the Protection of New Varieties of Plants* (UPOV Convention) (1961); *International Treaty on Plant Genetic Resources for Food and Agriculture* (ITPGRFA) (2001); TRIPS Agreement (*Agreement on Trade-Related Aspects of Intellectual Property Rights*)

Intellectual Property–Regulatory Complex: *acquis* for (*see* intellectual property *acquis* for agriculture); combined impact of IP and biosafety regimes, 4, 6, 15–16, 21; comple-

mentary mechanisms of market exclusion, 79–82; complexity not fully appreciated, 4, 34; definition, 4–5; difficulties created by, 32–33; effect on development of socially beneficial products, 85–87, 93–94; excessive intervention, 162; governance (*see* governance and the IP–Regulatory Complex); innovation and (*see* innovation incentives, socially beneficial products, and the IP–Regulatory Complex); judicial weakness in treating IP in these regimes, 162; little academic attention on impact on agricultural genomics, 16. *See also Convention on Biological Diversity* (CBD); *International Convention for the Protection of New Varieties of Plants* (UPOV Convention) (1961); *International Treaty on Plant Genetic Resources for Food and Agriculture* (ITPGRFA) (2001); TRIPS Agreement (*Agreement on Trade-Related Aspects of Intellectual Property Rights*)

Interdepartmental Committee on Biotechnology, 47

International Agricultural Research Centres (IARCs): ability to grant exclusive licences in certain cases, 158; description, 153; Easy-SMTA tool, 155; relationship with ITPGRFA, 155; seed banks held in trust for ITPGRFA, 145, 152–54. *See also* CGIAR Consortium; Consultative Group for International Agricultural Research (CGIAR)

International Convention for the Protection of New Varieties of Plants (UPOV Convention) (1961): breeders' rights at expense of farmers, 22–23, 146; cooperation with ITPGRFA, 161; developed countries' optimization of techno-scientific advantage, 147; farmers' privilege, not right, to save and replant seed, 23, 31, 35n7,

160, 167n15; limits on protection (scope, public interest, exhaustion), 224; members, 217; model for plant variety protection regimes, 23; plant breeders' rights, 146; plant "variety," characteristics defined (new, distinct, uniform, stable), 10n5; plant variety protection regimes in many countries, 5
International Crops Research Institute for the Semi-Arid Tropics (ICRISAT) (India), 92
International Maize and Wheat Improvement Center, 93
International Treaty on Plant Genetic Resources for Food and Agriculture (ITPGRFA) (2001), access to plant genetic resources, 22, 24–25, 144–45, 148, 149, ix; ambiguity re IP rights, 150, 155, 163–64; collaboration with global actors, 161, 164; contracting countries, ix; equitable sharing of benefits, 144–45, 149–50; farmers' rights as goal, 145, 148–49; funding through Benefit Sharing Fund, 150–52, 155, 163; funding of research projects, 152; "genetic resource commons," 25; goals same as CBD's, 147–48; guaranteeing sharing of plant genetic resources, ix, 22, 24–25; High Level Round Tables on the Treaty, 163, 165; implementation, 162; IP protection tightened before ITPGRFA, 145–47; judicial weakness in treating IP, 162; jurisdiction over ex situ seed banks (through CGIAR and IARCs) (Article 15), 145, 152–54; jurisdictional overlaps with CBD and CGIAR, 160–61; priorities set by FAO's Global Plan of Action, 149, 152, 161, 166n5; proprietary rights important for fostering innovation, 6; Rio+20 Six-Point Action Plan, 163, 165; source of germplasm important re rights, 6; Standard Material Transfer Agreement (SMTA), 25, 151, 153, 155, 166n9. *See also* CGIAR Consortium; Consultative Group for International Agricultural Research (CGIAR); *International Convention for the Protection of New Varieties of Plants* (UPOV Convention) (1961); TRIPS Agreement (*Agreement on Trade-Related Aspects of Intellectual Property Rights*)
IP. *See entries beginning with* intellectual property
ITPGRFA. *See International Treaty on Plant Genetic Resources for Food and Agriculture* (ITPGRFA) (2001)

Jain, Sonu, 108
Japan: both patent and plant variety protection available, 30; preference for GMO-free food, 116, 119
Jayaraman, K.S., 108
Johnson, James, 203
Joshi, Sharad: on farmer suicides in India, 104; push for immediate approval of transgenic cotton, 106; "Robin Hood" construction of stealth seeds, 104, 107, 108; on transgenic anti-bollworm crops in Gujarat, 104–5

Kalous, M.J., 46
Keraleeyan (peasant leader in India), 102
Kevles, Dan, 212
Kingsbury, Benedict, 220–21
Kneen, Brewster, 46
Krisch, Nico, 220–21

Lawrence, Thomas, 174
Leading into the Next Millennium (National Biotechnology Advisory Committee), 50
legitimization of innovations (cognitive, socio-political), 176–78, 186–87
Leibenstein, Harv, 175
Liaison Group of Biodiversity-Related Conventions, 161

life-cycle analysis of socially beneficial products, 69, 94n1
LMOs (living modified organisms), 20–21, 34n1, 35n4–5
Lobel, Orly, 200–1

Mahalingappa, Shankarikoppa, 8, 111–12, 127
Mahyco company: complaint re stealth anti-bollworm crops in Gujarat, 105; denial of development of "terminator seeds," 103–4; market exclusivity in India for *Bt* insect-resistant crops despite no patent, 80–81; MECH-915 variety of transgenic seeds, 108; official approval of *Bt* cotton, 107; resistance development in cotton not a problem, 126
Marden, Emily, 156
Martin, Michael, 177
Mayo Collaborative Serv. v. Prometheus Laboratories, Inc., 215
Mehta, N.P., 109
Menkel-Meadow, Carrie, 204–5
Modi, Narendra, 106
Monsanto: biotechnology (transgenic) leader in agriculture, 179, 180, 184; *Bt* corn producer, 74; cognitive legitimacy main focus in IP management, 180; Cry1Ac gene in *Bt* cotton, 105, 110, 113, 114; deferring application to release Roundup Ready wheat in India, 57; drought-tolerant corn, transgenic, 83; on farmers' rights re seed saving, 183–84; IP management approach and consequences, 180–84, 185, 187; market exclusivity for glyphosate-tolerant soybeans due to regulatory system, 81–82; market exclusivity in India for *Bt* insect-resistant crops despite no patent, 80–81; MECH-915 variety of transgenic seeds, 108; "Monsanto/terminator/suicide-seed" narrative in India, 102–4; official approval of *Bt* cotton in India, 107; portrayed as coercer of

farmers, 113–14; research on greater nutritional content in crops, 84; Roundup Ready 2 product, 81; Roundup Ready transgenic seeds, 180; socio-political concerns re Roundup, 181–82; socio-political legitimacy eroded in India, 183–84; target in Brazil for opposition to transgenic soy, 115, 117, 182–83; Water Efficient Maize for Africa (WEMA) initiative, 93. *See also* stealth seeds in Brazil; stealth seeds in India
Monsanto Canada Inc. v. Schmeiser (2004): component of plant patentable, 30, 216; inconsistency noted between *Patent Act* and *Plant Breeders' Rights Act*, 31; liability possible for the farmer, 55, 62n7; social acceptability of technology affected, 181
Monsanto Tech. LLC v. Cefetra BV, 215–16
Morrissey, Brian, 48–49

Nagoya Protocol (2014), 161, 164, 217
Nakagaki, Paul, 186
Naseem, Anwar, 122
National Biotechnology Advisory Committee (NBAC) (Canada), 45–46, 50
National Farmers Union, 54, 55
Nature Biotechnology, on regulatory system in India, 111
Navbharat 151, 105, 107, 108, 109, 114, 124, 128, 129n10, 131n23
Neofederalist Vision of TRIPS: The Resilience of the International Intellectual Property Regime (Dinwoodie and Dreyfuss), 212
new governance. *See* governance and the IP–Regulatory Complex
Nidera (Argentina), 118
North, Douglass, 174, 175

Operation Cremate Monsanto: activists in India burning *Bt* cotton trials, 103,

111, 112–13, 127; dissent of farmers, 8, 111–12

Organisation for Economic Co-operation and Development (OECD), 45

Paarlberg, R.L., 121

Paris Convention, 216–17, 223

Patel, Piyush, 108

Patent Act (Canada): applicability to plants uncertain, 30–31; court cases on patentability for plants, 30–31, 114, 181, 216; inconsistency with *Plant Breeders' Rights Act*, 31; IP protection regime, 17; no provision re farmers' right to save and reuse seed, 31; silent as to whether a plant is patentable, 30

patents: biotechnology clause in TRIPS Agreement, 23–24, 146–47, 165n3, 217; breadth of gene patents, 214–16; Canadian court cases on patentability for plants, 30–31, 114; control of socially beneficial technology extended beyond life of patent, 86; efficacy for product *vs.* process innovations, 173; encouragement and danger for innovation, 16, 75–76, 79, 174–75, 213, 224, ix; invalidation of patents on genomic DNA, 224–25; market exclusivity with no patent protection, 80–82; minimum standards for patent protection under TRIPS Agreement, 23; NBER study on value, 173–74; *Plant Patent Act* (US), 30, 212; *Plant Variety Protection Act* (US), 212, 214; possibility for sunflower genomics project (Canada), 33; tightening criteria to maximize innovation, 88; utility and plant patents both available in US, 30. *See also Patent Act* (Canada); *Plant Breeders' Rights Act* (Canada)

Patil, Rohidas, 106

Pawar, Sharad, 107

Pest Control Products Act (Canada), 48

Pharmaceutical Research and Manufacturers of America, 79

Phillips, Nelson, 174

Pioneer Hi-Bred (*later* Dupont Pioneer), 74

Pioneer Hi-Bred Ltd. v. Canada (1989), 30

Pisano, Gary, 176–77

Plant Biosafety Office (Canadian), 26

Plant Breeders' Rights Act (Canada): breeders' (research) exemption, 29, 31–32, 36n16; farmers' privilege, 29, 31; inconsistence with *Patent Act*, 31; IP protection regime for new plants, 17, 28–29

Plant Patent Act (US), 30, 212

Plant Variety Protection Act (US), 212, 214

Plants with Novel Traits (PNT) regime (Canada): applicability to sunflower project, 33, 37n18; criticism of PNT regulatory approach, 27; part of IP–Regulatory Complex, 4, 17; plants developed in Canada subject to PNT regime, 26; product rather than process analyzed, 26–27; products intended for consumption as food, 35n11

Platform for Translational Research on Transgenic Crops (PTTC) (India), 92

Poddar, Urbashi, 120–21

Prairie Registration Recommending Committee for Grain, 56

Pray, Carl, 122

Prudente, Antonio, 116

Public Intellectual Property Resource for Agriculture (PIPRA) (California), 92

Qaim, M., 122

Radio Telefis Fireann & Indep. Television Publ'ns Ltd. v. Comm'n of the European Communities, 216

Rana, Kashiram, 106

Rank Hovis (UK flour mill), 54

Ravetz, Jerome, 199

regulation of agricultural biotechnology, Canadian: applicable to all products with "novel trait," including genomics, 43, 60, 96n21; "biotechnology clause" in TRIPS Agreement, 23–24, 146–47, 165n3, 217; Canada party to CBD and WTO regimes, 26; Canadian Agricultural Research Council on regulatory framework, 44–45; Canadian Biotechnology Advisory Committee (CBAC), 44, 50–51, 53, 60; early views of risks of biotechnology, 44–45; efforts to minimize regulatory burdens, 19; Federal Framework for Biotechnology (1993) guiding principles, 47; genomics, proteomics, and cloning included, 43; GM crops, debate about, 43; Interdepartmental Committee on Biotechnology, 47; National Biotechnology Advisory Committee (NBAC), 45–46, 50; Plant Biosafety Office, 27; *Plant Breeders' Rights Act*, 17, 28–29, 31–32, 36n16; Plants with Novel Traits (PNT) regime, 4, 17, 26–27, 33, 37n18; report by HoC Committee on Agriculture and Agri-Food, 49–50, 53, 55, 56–57; report by HoC Committee on the Environment and Sustainable Development, 49; reports influencing frameworks (late 1980s), 46; science-based only, 43–49, 59–60, 61–62; scientism and, 44, 58–59; *Seeds Act*, 26, 48, 56; social and ethical concerns not considered, 43–46, 47–49, 57–59, 60–62; Workshop on Regulating Agricultural Products of Biotechnology (1993), 48. *See also* agricultural biotechnology, social and ethical concerns; Canadian Food Inspection Agency (CFIA); regulation of agricultural biotechnology, international

regulation of agricultural biotechnology, international: basic goal of biosafety, 77; Goldilocks ("just right" amount) regulation, 126–27; incentives and (*see* innovation incentives, socially beneficial products, and the IP–Regulatory Complex); international impact of national policies due to globalization, 78–79; national policies based on national priorities, 26; needed to encourage innovation, ix; Responsible Research and Innovation (RRI) concept, 61; safety of product, proof the responsibility of innovator, 77; scope of approval determined on case-by-case basis, 78; SPS Agreement (*Agreement on the Application of Sanitary and Phytosanitary Measures*), 19–20, 34n3; time-limited approvals, 78; WTO's *General Agreement on Tariffs and Trade* (GATT), 19–20. *See also Convention on Biological Diversity* (CBD); *International Convention for the Protection of New Varieties of Plants* (UPOV Convention) (1961); *International Treaty on Plant Genetic Resources for Food and Agriculture* (ITPGRFA) (2001); TRIPS Agreement (*Agreement on Trade-Related Aspects of Intellectual Property Rights*)

Regulation of Plant Biotechnology in Canada (Caldwell and Duke), 46

Regulation of Plant Biotechnology in Canada: Part 2, The Environmental Release of Genetically Altered Plant Material (Kalous and Duke), 46

Research Foundation for Science, Technology and Ecology (RFSTE), 113–14

Responsible Research and Innovation (RRI), 61

Rio+20 Six-Point Action Plan (UN, 2012), 163, 165

Rodrigues, Roberto, 120

Roundup Ready–tolerant crops:
Monsanto's IP management of
Roundup Ready transgenic seeds,
180–81; regulatory system control
like IP control, 81–82; release of
Roundup Ready wheat delayed in
India by Monsanto, 57; Roundup
Ready 2 product, 81, 86; Schmeiser
case (2004) re plant patentability,
30, 31, 55, 181, 216; social benefits
of, 74; socio-political concerns, 181–
82; soybeans, 81–82, 86, 116–17,
118; tolerance genetically engineered, 71, 72, 75, 81–82
Royal Society of Canada, 44, 50
Royal Society of Canada Expert Panel,
51–52
Rutala, Purshottam, 106

Sampaio, Maria José, 121
Sanders, Lynn, 203
Saskatchewan Association of Rural
Municipalities, 55–56
Saskatchewan Organic Directorate, 54,
62n7
Schmeiser, Percy, 30, 31, 55, 62n7, 114,
181, 216
scientism, 44, 58, 59
Scotts Company LLC, 84
seed banks (ex situ), 144–45, 152–54.
See also CGIAR Consortium;
Consultative Group for International
Agricultural Research (CGIAR);
*International Treaty on Plant
Genetic Resources for Food and
Agriculture* (ITPGRFA)
seeds: defined in *Seeds Act*, 26; farmers'
privilege (to save and replant seed),
23, 29; hybridization and proprietary protection, 5, 10n3, 21; property rights dependent on monitoring
by technology, 123; risks and regulation of recombinant-DNA technology, 124–26; seed saving permitted
for most of recent history, 5, 10n2;
social cost-benefit analysis of

recombinant-DNA technology, 124–
28; stealth seeds, common factors,
122; stealth seeds and dissension in
coalitions for poor, 8, 111–13, 118,
120–21, 123; stealth seeds spread
beyond Brazil and India, 122. *See
also* seed banks (ex situ); stealth
seeds in Brazil; stealth seeds in India
Seeds Act (Canada), 26, 48, 56. *See
also* Plants with Novel Traits (PNT)
regime (Canada)
Seeds of Suicide (Shiva et al.), 103
Sharma, Manju, 106
Sharma, R.P., 108
Shetkari Sanghatana, 104
Shiva, Vandana, 103, 113–14
Shuen, Amy, 176–77
Silva, Marina, 120
Singh, Ajit, 106, 128
Slade, Alison, 221
social and ethical concerns. *See* agricultural biotechnology, social and
ethical concerns
socially beneficial products: biofuel
crops (orphan crop), 75; commercially successful products more of
interest to private sector, 143–44;
common public perceptions and as-
sessments, 68, 69–70; control of
technology extended beyond life of
patent, 86; control of technology
platforms by oligopolies reinforced,
85–86; development and the IP–
Regulatory Complex, 85–87, 93–94;
drought and stress tolerance (orphan
traits/crops), 75, 82–83, 87; economic measures of costs/benefits,
68, 69; entrepreneurship discouraged with industry consolidation,
86–87; HarvestPlus (HP) program of
CGIAR, 154; herbicide tolerance (*see*
Roundup Ready–tolerant crops);
incentives, improvement, 93–94;
incentives *not* to innovate, 86; innovation skewed to applications for
large markets, 85; IP and (*see*

International Treaty on Plant Genetic Resources for Food and Agriculture [ITPGRFA]); IP regimes as threat to development of, 148; life-cycle analysis, 69, 94n1; nutritional content improvement (orphan trait/crop), 75, 84–85, 154; orphan traits/crops, 75, 82–85, 87, 154; shift to nontransgenic approaches to improve crop genetics, 87; TRIPS Agreement. *See also* innovation incentives, socially beneficial products, and the IP–Regulatory Complex

socio-political legitimization of innovations, 176–78

SPS Agreement (*Agreement on the Application of Sanitary and Phytosanitary Measures*), 19–20, 34n3

Standard Material Transfer Agreement (SMTA) (of ITPGRFA), 25, 151, 153, 155, 166n9

stealth seeds in Brazil: advantages seen by farmers, 120–21; corporate globalization a target for opposition, 117, 119, 121; CTNBio federal agency responsible for biosafety, 116; disconnect between farmers and organizations claiming representative status, 8, 117–18, 120–21, 123; disjointed regulatory environment for transgenics, 118; future of such seeds in legal limbo, 121; GMO-free soy threatened by stealth soy, 119; governments' differing support for transgenic soy, 117, 119–20, 183; legitimacy to rule on transgenics contested, 116–17; Monsanto a target for opposition, 115, 117, 182–83; official support but challenges from civil society, 115–18, 120; post-fact regularization of these seeds, 120; property rights dependent on monitoring by technology, 123; public researcher EMBRAPA's new transgenic soy, 121, 182–83; regulations irrelevant due to stealth transgenic

seeds, 115; underground seed market flourishing, 8, 115, 117–18, 120–21; widespread planting of transgenic soy, 117–18, 120

stealth seeds in India: activists in India burning *Bt* cotton trials, 103, 111, 112–13, 127; comparison of MECH-162 with non-*Bt* hybrid, 110–11; competition for official anti-bollworm varieties, 107; Cry1Ac gene in *Bt* cotton, 105, 110, 113, 114; disconnect between farmers and organizations claiming representative status, 8, 111–13, 123; farmer organization (DKSS) to fight globalization, 102–3; farmer organization (KRRS) to fight transgenic crops, 111–12; farmers' adoption of both Monsanto and public sector versions of *Bt* cotton, 109–10; farmers' realization of biotechnology benefits, 112–13, 114–15; farmers' seed saving/exchange/ experimentation, 111; legal *vs.* illegal packets of transgenic seeds, 108; MECH-915 variety (Mahyco-Monsanto), 108; "Monsanto/terminator/suicide-seed" narrative, 102–4; nationalist argument against transgenic crops, 113–14, 183; *Navbharat variants* (of 151) bred by farmers, 107–8, 124, 128; official approval of *Bt* cotton in India (2002), 107, 127; official seed sales increasing, 113; Operation Cremate Monsanto, 103, 111, 112–13, 127; performance variation in *Bt* cotton and "failure" reports, 110, 113; property rights dependent on monitoring by technology, 123; push for immediate approval of transgenic cotton, 106–7; "Robin Hood" construction of stealth seeds, 104, 107, 108; socio-political legitimacy of Monsanto eroded, 183–84; transgenic anti-bollworm seed (Navbharat 151) in Gujarat, 104–6, 124; underground

seed market flourishing, 8, 107–9; weak institutions as enabling underground trade, 184
Stewart, Richard, 220–21
sunflower (*Helianthus* sp.), 17–18
sunflower genomics project (Canada), 16, 18, 33, 37n18
Supreme Court of Canada: *Harvard College v. Canada* (2002), 30, 216; *Monsanto Canada Inc. v. Schmeiser* (2004), 30–31, 55, 62n7, 181, 216
Swaminathan, M.S., 111
Syngenta Ltd., 74, 83, 84
Szerb, Laszlo, 175

Tait, Joyce, 180
Teece, David, 176–77
Trans-Pacific Partnership (TPP), 217–18
TRIPS Agreement (*Agreement on Trade-Related Aspects of Intellectual Property Rights*): biotechnology clause and patent protection for life forms, 23–24, 146–47, 165n3, 217; Brazil's legislation concerning IP, 116; compliance mandatory for WTO members, 147; controlling abuse of IP rights, 223; developed countries' optimization of techno-scientific advantage, 147; members, 146, 147; minimum standards for patent protection, 23; plant varieties' protection by patents, 23; scope limitations, 222; subject matter exclusions, 221; TRIPS-plus agreements due to uncertainties in national regimes, 24, 217–18
Trubek, David, 200–1
Trubek, Louise, 200–1

United Kingdom, human rights norms and IP law, 223
United Nations Food and Agriculture Organization (FAO): funding support for ITPGRFA, 152; on hunger worldwide, viii; management of ex situ seed banks, 145; priorities for ITPGRFA under Global Plan of Action, 149, 152, 161, 166n5; value of the sunflower as a crop, 18
United Nations' Rio+20 Six-Point Action Plan, 163, 165
United States: *Anti-Counterfeiting Trade Agreement* (US), 217; *Bilski v. Kappos* (US), 221–22; champion of TRIPS Agreement, 146; controlling abuse of IP rights, 223–24; criticism of expert-regulatory model of policy-making, 198; *Diamond v. Chakrabarty* (US), 35n8, 212–13; economic benefits of *Bt* insect-resistant crops, 72; *Ex parte Hibberd*, 30; genomics research programs, 15; *Harper & Row Publishers, Inc. v. Nation Enterprise* (US), 222–23; natural laws not patentable, 215; nontransgenic R&D methods due to costs, 87; *Plant Patent Act*, 30, 212; *Plant Variety Protection Act*, 212, 214; plants deemed to patentable subject matter, 5; public concern about agricultural biotechnology, 57; scope limitations to IP rights, 222; TRIPS-plus trade agreements with other countries, 217–18; utility and plant patents available, 30
Upadhyay, Labshankar, 114
UPOV Convention. *See International Convention for the Protection of New Varieties of Plants* (UPOV Convention) (1961)

Vanclief, Lyle, 53, 57
Vasconcellos, Ronaldo, 117
Vasudevan, P.A., 102
Vector Tobacco, 84, 96n23

Water Efficient Maize for Africa (WEMA) initiative, 93
Webb, Justin, 175
Western Canadian Wheat Growers Association, 54

wheat. *See* GMOs (genetically modified organisms)

Wield, David, 180

WIPO. See World Intellectual Property Organization (WIPO)

World Intellectual Property Organization (WIPO), 161, 217

World Trade Organization (WTO): compliance with TRIPS Agreement mandatory for WTO members, 147; Dispute Settlement Board for nutrition or agricultural research, 224; GATT (*General Agreement on Tariffs and Trade*), 19–20; sociopolitical concerns re WTO support for multinational companies, 181–82; SPS Agreement (*Agreement on the Application of Sanitary and Phytosanitary Measures*), 19–20, 34n3. *See also* TRIPS Agreement (Agreement on Trade-Related Aspects of Intellectual Property Rights)

Wright, Erik Olin, 202, 203

Wu, Felicia, 74

Wynne, Brian, 59

Printed and bound in Canada by Friesens

Set in Segoe and Warnock by Artegraphica Design Co. Ltd.

Copy editor: Frank Chow

Indexer: Patricia Buchanan